Essentials of Bioorganic Chemistry

Essentials of Bioorganic Chemistry

Edited by
Jeremy Riordan

Larsen & Keller
www.larsen-keller.com

Essentials of Bioorganic Chemistry
Edited by Jeremy Riordan
ISBN: 978-1-63549-674-1 (Hardback)

© 2018 Larsen & Keller

▤ Larsen & Keller

Published by Larsen and Keller Education,
5 Penn Plaza,
19th Floor,
New York, NY 10001, USA

Cataloging-in-Publication Data

Essentials of bioorganic chemistry / edited by Jeremy Riordan.
       p. cm.
Includes bibliographical references and index.
ISBN 978-1-63549-674-1
1. Bioorganic chemistry. 2. Biochemistry. 3. Chemistry, Organic. I. Riordan, Jeremy.
QP550 .E87 2018
572--dc23

For more information regarding Larsen and Keller Education and its products, please visit the publisher's website www.larsen-keller.com

Table of Contents

Preface

The study of using organic chemistry to understand and analyse the biological processes is referred to as bioorganic chemistry. It is used to analyse the kinetics, synthesis and structure of organic chemicals. The subject includes an in-depth study of cofactors, metalloenzymes, etc. Biophysical organic chemistry is a sub-part of bioorganic chemistry which deals with the study of molecules using the elements of organic chemistry. This book elucidates the concepts and innovative models around prospective developments with respect to bioorganic chemistry. Most of the topics introduced in it cover new techniques and the applications of the subject. This textbook will serve as a valuable source of reference for those interested in this field.

A detailed account of the significant topics covered in this book is provided below:

Chapter 1- Bioorganic chemistry is the study of organic chemical principals of biological processes. Various possibilities exist through the application of this field like the construction of artificial enzymes, synthetic polymers and synzymes. This chapter will provide an integrated understanding of bioorganic chemistry.

Chapter 2- Proximity allows bond polarization which further results in an increase in the rate of the reaction of enzyme substrate interactions. It also reduces entropy. This process improves biological reactions like ligations or addition reactions. This chapter provides a plethora of interdisciplinary topics for better comprehension of proximity effect in organic chemistry.

Chapter 3- Cell is the basic unit of all living organisms. It consists of a membrane with a cytoplasm within it. Cells form tissues and tissues collectively form organs. Cells are mainly of two types, prokaryotic and eukaryotic. The aspects elucidated in this chapter are of vital importance, and provide a better understanding of living cells.

Chapter 4- The chemical processes that occur within an organism can be studied under the subject of biochemistry. Organic reactions are chemical reactions which involve organic compounds. This chapter helps the readers in developing a better understanding of the biochemical and organic reaction.

Chapter 5- Chemical biology, as a subject, focuses on chemistry, physics and biology. It attempts to understand principles related to chemistry in order to apprehend biology and the functions of an organism. The section closely examines the key concepts of chemical biology to provide an extensive understanding of the subject.

Chapter 6- Amino acids are the building block of proteins. The specific pattern of an amino acid depend son the sequence of the protein. Organic molecules which consist of acidic carboxyl group and an organic R group are known as amino acids. The main elements of it are oxygen, hydrogen, nitrogen and carbon. This chapter has been carefully written to provide an easy understanding of the varied facets of amino acids.

Chapter 7- The small chains formed by amino acids are linked by bonds formed by peptides. Peptides are smaller in size than proteins and usually consist of 2 to 50 amino acids. The topics elaborated in this chapter will help in gaining a better perspective on the subject of peptide bond and synthesis.

It gives me an immense pleasure to thank our entire team for their efforts. Finally in the end, I would like to thank my family and colleagues who have been a great source of inspiration and support.

Editor

Fundamentals of Bioorganic Chemistry

Bioorganic chemistry is the study of organic chemical principals of biological processes. Various possibilities exist through the application of this field like the construction of artificial enzymes, synthetic polymers and synzymes. This chapter will provide an integrated understanding of bioorganic chemistry.

Bioorganic Chemistry

Bioorganic Chemistry can be defined as a branch of chemistry or broadly speaking a branch of science which utilizes the principles, tools and techniques of organic chemistry to the understanding of biochemical/biophysical process.

As for example, the classical chemistry of natural products with its characteristic triad of isolation, structural proof and total synthesis is an evident, but purely organic ancestor. Likewise, inquiry into the biosynthetic pathways for the same natural products is plain biochemistry. But when the total synthesis of a neutral product explicitly is based upon the known route of biosynthesis or if the biosynthesis has been translated into structural and mechanistic organic chemical language, one is clearly dealing with bioorganic chemistry.

Organic chemistry deals with:-Structure Design, synthesis, and kinetics (physical organic).

1. Structure Design: It guides us of how potential the interaction between structures and the biological partners.

2. Synthesis: Synthesis provides us with compounds which might be the analogue or the mimic of natural species and may not have created in sufficient quantity for investigation by nature.

3. Kinetics: Physical organic chemistry and analytical methodology provide quantitative measures and intimate details of reaction pathways.

Bioorganic chemistry is a rapidly growing scientific discipline that combines organic chemistry and biochemistry. While biochemistry aims at understanding biological processes using chemistry, bioorganic chemistry attempts to expand organic-chemical researches (that is, structures, synthesis, and kinetics) toward biology. When investigating metalloenzymes and cofactors, bioorganic chemistry overlaps bioinorganic chemistry. Biophysical organic chemistry is a term used when attempting to describe intimate details of molecular recognition by bioorganic chemistry.

Bioorganic chemistry is that branch of life science that deals with the study of biological processes using chemical methods.

Why the Term Bio-organic Chemistry?

As we discussed earlier, that the organic chemistry is related to the development of methodology to synthesize organic molecules of biological importance/analogues. However, not all the analogues are potent to have response to/or with biological molecules. So, modification of synthesis is necessary which is only possible from a thorough study of biological process, a part of biochemistry.

On the other hand, knowledge of biochemistry gives the idea of what would be useful to synthesis for a fruitful response which can only be possible via organic chemistry.

Therefore, the need for the multidisciplinary approach become obvious and there must have to have two laboratories-i) one for the synthesis and ii) another for the biological study. Thus, knowledge of organic chemistry give rise to the concept of building of organic models chemically synthesized in the laboratory to study the complex biological processes.

Bioorganic chemistry is thus, a young and rapidly growing science arising from the overlap of biochemistry and organic chemistry.

Bio-organic Chemistry-A Borderline Science-its Multiple Origin:

1. Enzyme Chemistry: For some hydrolytic enzymes the catalyzed reaction has been translated already into a series of normal organic reaction steps. At the same time organic chemists are mimicking the characteristics of enzyme catalysis in model organic reactions dealing with both the rate of reaction and specificity. Investigations, involving metalloenzymes and cofactors, the contiguous areas of bioorganic and bioinorganic chemistry also merge.

2. Nutritional Research: Knowledge of biochemistry enables us to recognize the factors essential in the human diet, and their structures and syntheses with the help of organic chemistry led to the recognition of the modes of action of the so-called vitamins and related cofactors, or coenzymes.

3. Hormone Research: Secreted factors that exert a stimulatory effect on cellular activity, the hormones, could be better understood at the molecular level once their structure determinations and syntheses made them available in reasonable amounts with the help of organic chemists.

4. Natural Products Chemistry: Concepts of the biogenesis of natural products played, and continues to play, a major role in the development of bioorganic chemistry. The classical chemistry of natural products with its characteristic triad of isolation, structural proof and total synthesis is an evident, but is a purely organic ancestor. Likewise, inquiry into the biosynthetic pathways for the same natural products is plain biochemistry. But when the total synthesis of a natural product explicitly is based upon the known route of biosynthesis or if the biosynthesis has been translated into structural and mechanistic organic chemical language, one is clearly dealing with bioorganic chemistry.

5. Molecular Recognition: The term molecular recognition refers to the specific interaction between two or more molecules through non-covalent bonding such as hydrogen bonding,

metal coordination, hydrophobic forces, van deer Waals forces, pi-pi interactions, electrostatic and/or electromagnetic effects and is purely physical organic chemistry origin. The host and guest involved in molecular recognition exhibit molecular complementarities. Molecular recognition plays an important role in biological systems and is observed in between receptor-ligand, antigen-antibody, DNA-protein, sugar-lectin, RNA-ribosome, etc. An important example of molecular recognition is the antibiotic vancomycin that selectively binds with the peptides with terminal D-alanyl-D-alanine in bacterial cells through five hydrogen bonds. The vancomycin is lethal to the bacteria since once it has bound to these particular peptides they are unable to be used to construct the bacteria's cell wall. Therefore, the composite term, biophysical organic chemistry, has been used as a detailed descriptor in molecular recognition.

6. Protein Chemistry (sequencing) vs. Application of Reagents: A simple chemical applied according to a well recognized concept can be responsible for a great advance in biological chemistry. Thus, through the reaction of cyanogens bromide, Bernhard Witkop translated neighbouring group participation into selective, limited, non-enzymatic cleavage at methionine in a peptide chain.

7. Reagents vs. Modern Biotechnology: Application of the reagent has aided not only the correct sequencing of peptide segments of many proteins but also the production, through genetic engineering, of human insulin by means of a methionyl-containing precursor version at each step provides the basis of modern biotechnology: the, automated synthesis of polypeptide and polynucleotide chains and the sequencing of DNA and RNA.

Therefore, as organic chemists, we use chemically-based biotechnology and continue to add other techniques that are not only applicable but in some cases requisite: fluorescence sorting and probing; recombinant DNA technology; cloning; plasmid construction. Organic chemistry approaching 100% con-combinatory procedures; the polymerase chain reaction (PCR); all of the latest separation and spectroscopic methodology with computer analysis; and the generous use -- as reagents -- of bacteria, fungi, enzymes, whole cells, and ground liver microsomes, inter alia.

Artificial Enzyme

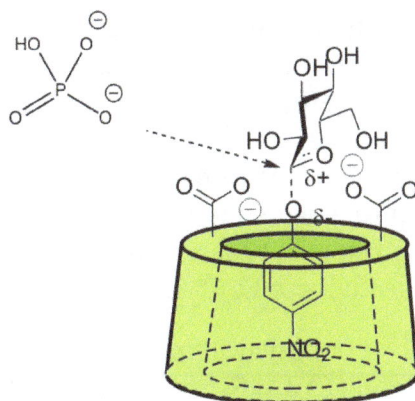

Schematic drawing of artificial phosphorylase

An artificial enzyme is a synthetic, organic molecule or ion that recreate some function of an enzyme. The area promises to deliver catalysis at rates and selectivity observed in many enzymes.

History

Enzyme catalysis of chemical reactions occur with high selectivity and rate. The substrate is activated in a small part of the enzyme's macromolecule called the active site. There, the binding of a substrate close to functional groups in the enzyme causes catalysis by so-called proximity effects. It is possible to create similar catalysts from small molecule by combining substrate-binding with catalytic functional groups. Classically artificial enzymes bind substrates using receptors such as cyclodextrin, crown ethers, and calixarene.

Artificial enzymes based on amino acids or peptides as characteristic molecular moieties have expanded the field of artificial enzymes or enzyme mimics. For instance, scaffolded histidine residues mimics certain metalloproteins and -enzymes such as hemocyanin, tyrosinase, and catechol oxidase.

In December 2014, it was announced that active enzymes had been produced that were made from artificial molecules which do not occur anywhere in nature.

Nanozymes

Nanozymes are nanomaterials with enzyme-like characteristics. They have been widely explored for various applications, such as biosensing, bioimaging, tumor diagnosis and therapy, antibiofouling. In 1996 and 1997, Dugan et al. discovered the superoxide dismutase (SOD) mimicking activiteis of fullerene derivatives. In 2004, the term "nanozymes" was coined by Flavio Manea, Florence Bodar Houillon, Lucia Pasquato, and Paolo Scrimin. In 2006, nanoceria (i.e., CeO_2 nanoparticles) was used for preventing retinal degeneration induced by intracellular peroxides. In 2007, Xiyun Yan and coworkers reported that ferromagnetic nanoparticles possessed intrinsic peroxidase-like activity. In 2008, Hui Wei and Erkang Wang developed an iron oxide nanozyme based sensing platform for bioactive molecules (such as hydrogen peroxide and glucose). In 2012, recombinant human heavy-chain ferritin coated iron oxide nanoparticle with peroxidase-like activity was prepared and used for targeting and visualizing tumour tissues. In 2012, vanadium pentoxide nanoparticles with vanadium haloperoxidase mimicking activities were used for preventing marine biofouling. In 2014, it was demonstrated that carboxyfullerene could be used to treat neuroprotection postinjury in Parkinsonian nonhuman primates. In 2015, a supramolecular regulation strategy was proposed to modulate the activity of gold-based nanozymes for imaging and therapeutic applications. A nanozyme-strip for rapid local diagnosis of Ebola was developed. An integrated nanozyme has been developed for real time monitoring the dynamic changes of cerebral glucose in living brains. A book entitled "Nanozymes: Next Wave of Artificial Enzymes" was published. Oxidase-like nanoceria has been used for developing self-regulated bioassays. Histidine was used to modulate iron oxide nanoparticles' peroxidase mimicking activities. Gold nanoparticles' peroxidase mimicking activities were modulated via a supramolecular strategy for cascade reactions. A molecular imprinting strategy was developed to improve the selectivity of Fe_3O_4 nanozymes with peroxidase-like activity.

Several conferences have focused on nanozymes. In 2015, a nanozyme workshop for was held at the 9th Asian Biophysics Associatation (ABA) Symposium. In Pittcon 2016, a Networking entitled "Nanozymes in Analytical Chemistry and Beyond" was devoted to nanozymes.

Artificial Enzyme for Transamination

Nature taught us how to synthesize amino acids by using pyridoxamine phosphate coenzyme to perform a transamination with a keto acid. Therefore, R. Breslow and others have studied models for such reactions. Thus, they have constructed a model enzyme system for the process by using a hydrophobic derivative of pyridoxamine as the coenzyme mimic and a polyamine with an added hydrophobic core as the enzyme mimic. The coenzyme bound into this non-polar region, and the substrates as well bound into it, especially if they carried hydrophobic groups, as in the keto acid that formed DL-alanine/phenylalanine and other enantiomerically pure L-aminoacids.

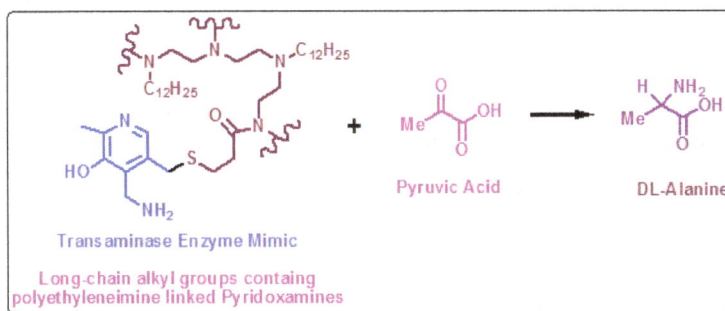

Polyethyleneimine linked pyridoxamine as artificial transaminase enzyme mimic

The amination of ketoacids to amino acids by pyridoxamine is greatly accelerated when the pyridoxamine is covalently linked to polyethylenimine carrying N-methyl and N-lauryl groups. The polyamine catalyzes the reaction using acid and base groups, the lauryl groups increase k_2 by producing a nonpolar medium in which the reaction occurs, and the lauryl groups promote binding of hydrophobic substrates. The result is that the amination of indolepyruvic acid to produce tryptophan is accelerated by 240000-fold.

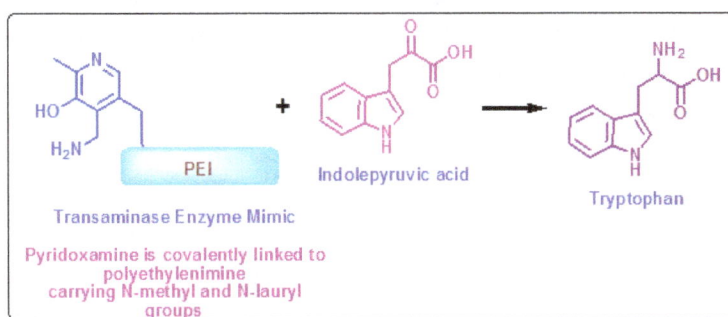

Polyethyleneimine carrying N-methyl and N-lauryl groups linked pyridoxamine as artificial transaminase enzyme mimic.

PAMAM dendrimers from generations 1–6 were synthesized with pyridoxamine in their core. They transaminated pyruvic and phenylpyruvic acids in water to alanine and phenylalanine, respectively, with Michaelis–Menten kinetics and high effectiveness compared with simple pyridoxamine. The largest dendrimerssimilar in size to some globular proteinswere comparable in effectiveness to a previous polyethylenimine (PEI)–pyridoxamine catalyst, and to a protein–pyridoxamine catalyst, but not as effective as a previous PEI–pyridoxamine carrying lauryl hydrophobic groups.

The new catalysts showed both general acid/base catalysis by their amino groups and hydrophobic binding of the phenylpyruvate substrate.

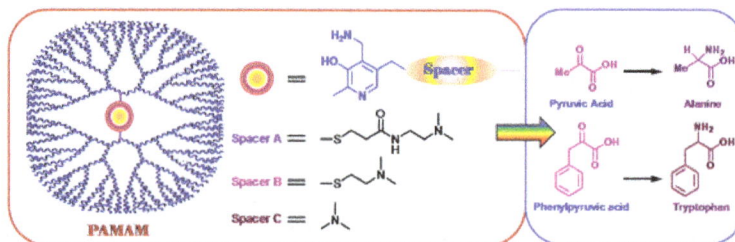

PAMAM dendrimers linked pyridoxamine catalyst as artificial transaminase enzyme mimic.

Isotactic polyethylenimines with (S)-benzyl side chains were synthesized from 4-(S)-4-benzyl-2-oxazolines. When α-keto acids were subjected to transamination in the presence of this polymer, and a pyridoxamine coenzyme modified with hydrophobic chains, enantioselectivity toward the natural isomer (l > d) was observed, followed by racemization of the amino acid products. However, the racemization did not occur when the coenzyme was covalently attached to the polymer.

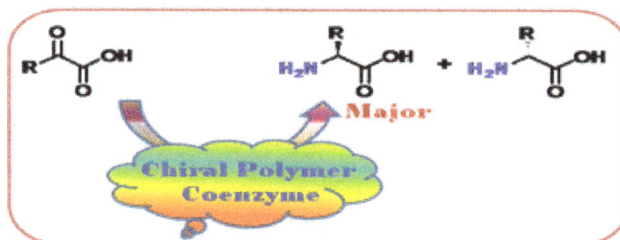

Isotactic polyethylenimines induce formation of l-Amino Acids in transamination.

Natural enzymes are macromolecules, but most enzyme models are small molecules. To mimic the role of the macromolecular character of enzymes in catalysis, we have recently studied some polymeric and dendrimeric enzyme models. We found a great increase of transamination rate for the pyridoxamine/ketoacid system when we covalently linked pyridoxamine to polyethylenimine (PEI) carrying some attached lauryl groups, or covalently located one pyridoxamine unit at the core of poly(amidoamine) (PAMAM) dendrimers.

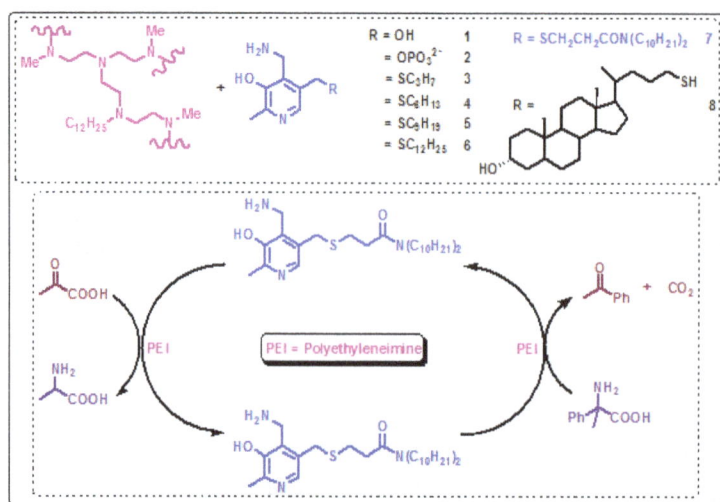

Noncovalent polymer-pyridoxamine systems as better transaminase mimics

In our polymeric and dendrimeric mimics the pyridoxamine cofactor was covalently attached to PEI or PAMAM. However, in the real transaminases the pyridoxamine cofactor forms a noncovalent complex with the enzyme protein matrix. Thus we have now developed some noncovalent polymer-pyridoxamine systems as better transaminase mimics, in which the coenzyme reversibly binds into the polymer. We find that they are even more potent than the covalently linked analogues, since they bind into the hydrophobic region of the polymer. Furthermore, we have now developed a novel catalytic cycle that recycles the pyridoxal cofactor to the pyridoxamine, and for the first time achieves high turnovers in transamination in such enzyme mimics.

The transaminase activity of two new semi-synthetic RNase-S proteins incorporating a pyridoxamine moiety at the active site has been evaluated. A chemically competent derivative of pyridoxamine phosphate was incorporated into the C-peptide fragments of these non-covalent protein complexes in the form of an unnatural coenzyme-amino acid chimera, 'Pam'. The chimeric Pam residue integrates the heterocyclic functionality of pyridoxamine phosphate into the side chain of an alpha-amino acid and was introduced instead of Phe8 into the C-peptide sequence via standard solid phase methodology. The two semi-synthetic Pam-RNase constructs were designed to probe whether the native ribonuclease catalytic machinery could be enlisted to modulate a pyridoxamine-dependent transamination reaction. Both RNase complexes, H1SP and S1SP, exhibited modest rate enhancements in the Cu(II)-assisted transamination of pyruvate to alanine under single turnover conditions, relative to 5'-deoxypyridoxamine and the uncomplexed C-peptide fragments. Furthermore, multiple turnovers of substrates were achieved in the presence of added L-phenylalanine due to recycling of the pyridoxamine moiety. The modest chiral inductions observed in the catalytic production of alanine and the differences in reactivity between the two proteins could be rationalized by the participation of a general base (His12) in complex H1SP, and by the increased tolerance for large amino acid substrates by complex S1SP, which contains serine at this position. The pyridoxamine-amino acid chimera will be useful in the future for examining the coenzyme structure/ function relationships in a native-like peptidyl architecture.

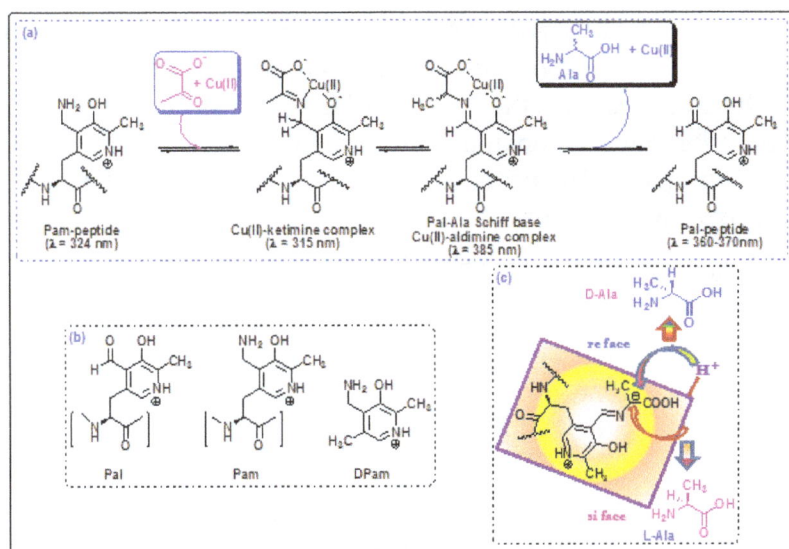

(a) Structures of intermediates in the Cu(II)-assisted transamination of pyruvate to alanine by Pam-containing peptides; (b)Structures of the pyridoxal (Pal) and pyridoxamine (Pam) coenzyme–amino acid chimeras, and deoxypyridoxamine (DPam); (c)Enantioselective production of alanine in the transamination of pyruvate, due to selective protonation of a particular face of the Pam-aldimine carbanion intermediate.

Models for Nicotinamide Dehydrogenase Reactions

Although enzymatic dehydrogenation reactions involving nicotinamide coenzymes have been studied extensively, there remains considerable controversy as to the molecular mechanism of hydrogen transfer in these reactions. The overall reaction accomplishes direct hydrogen transfer between the substrate and the 4-position of the coenzyme's nicotinamide ring.

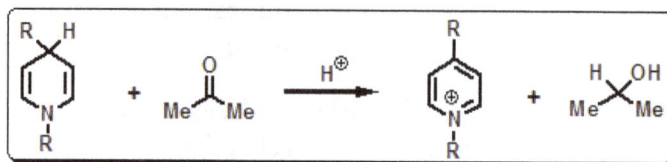

Nicotinamide coenzymes mediated H-transfer reaction

Nicotinamide coenzymes are involved in enzymatic reactions that interconvert alcohols, amines or activated methylene compounds with ketones, imines or olefins respectively. The chemistry of these interconversions tends to be very rapid. In fact, in some cases the rates of hydrogen transfer surpass those of product release from the enzyme. Thus far, three types of mechanisms have been proposed for hydrogen transfer in these reactions:

1. Direct bimolecular hydride transfer, in which the hydrogen nucleus and both electrons are transferred as a single unit.

2. Free radical mechanisms in which the hydrogen is transferred as two hydrogen atoms or as electrons and protons in separate steps.

3. Mechanisms involving covalent intermediates in which the coenzyme and substrate become covalently linked and the electrons are transferred through the covalent bonds while the hydrogen is transferred as a proton.

Each of these mechanisms has its proponents, however, none is generally accepted. The hydride and radical mechanisms can be ruled out on the basis of the facility of enzyme reactions. Pathways involving covalent intermediates solve the problem of high activation energies by employing equilibrium controlled addition-elimination and proton transfer reactions. The mechanism proposed by Hamilton is shown in Figure. The proposed mechanism is very similar to retro-ene reactions that are known to occur with allyl ethers.

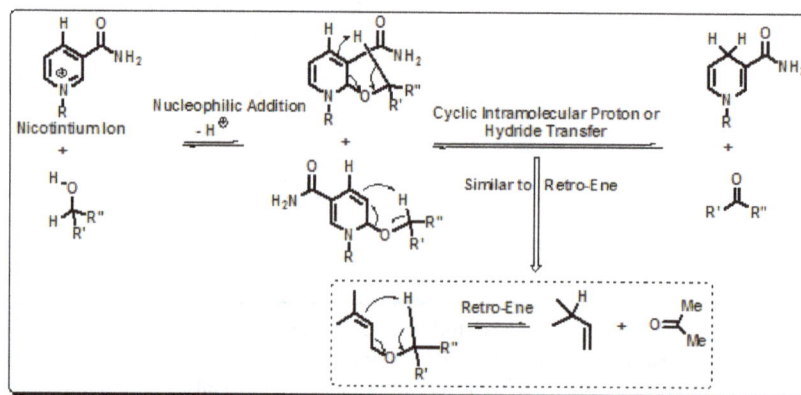

Proton transfer and oxidation of alcohol to ketone by Nicotinium ion.

Bioreductants and their Inspired Organoreductants: Chemical Mechanisms Relevant to Catalysis (e.g. NADH)

Some Examples of Bioreductant:

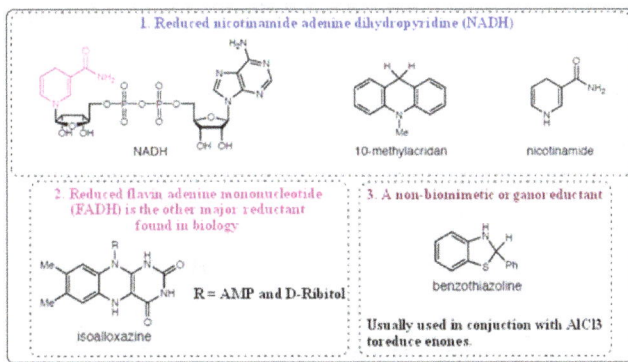

Some examples of bioreductants.

A. Hantzsch Ester (Hantzsch dihydropyridine)- a NADH Model:

1. Hantzsch esters reduce imine derivatives.

2. Hantzsch esters effectively reduce electrophilic olefins.

3. Nitro- and carbonyl alkene reductions

    a. β-nitroalkene derivatives

    b. α-β-Unsaturated aldehydes and ketones

4. Lewis acid catalyzed Hantzsch reactions

    a. Olefin reduction with SiO$_2$

    b. Reductive amination using Sc(OTf)$_3$ and LiClO$_4$

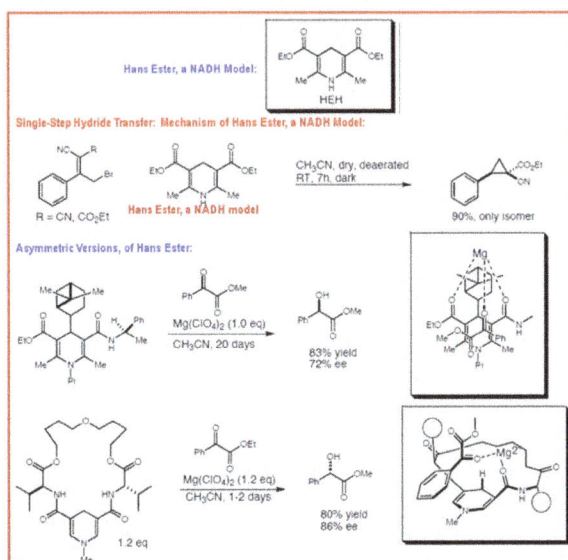

Hantzsch ester (Hantzsch dihydropyridine)- a NADH Model and its reactions.

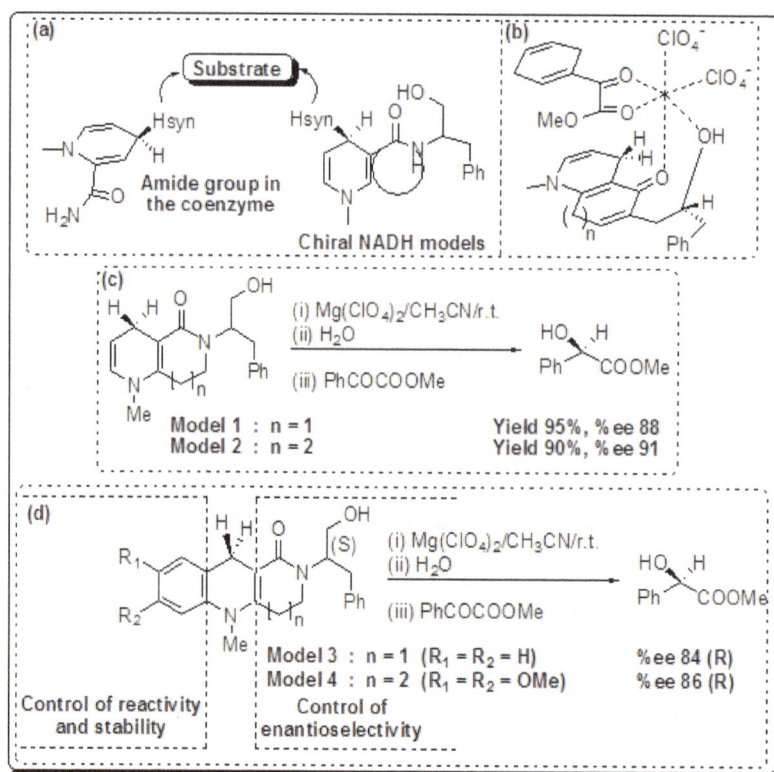

(a) Conformational analysis of the amide group in the coenzyme and biomimetic approach in the design of new chiral NADH models; (b) Proposed ternary complex (model/Mg^{2+}/substrate) of models bearing aminoalcohol as chiral auxiliary; (c) enantioselectivity during the reduction of methyl benzoylformate with model 1 and 2; (d) asymmetric reduction of methyl benzoylformate with model 3 and 4.

B.   Biomimetic NADH Models: "Nucleophile-Transferring Agents"

Biomimetic NADH models have been almost exclusively explored to develop redox processes. To extent the potential of these biomimetic tools, Vincent Levacher et al. probed their aptitude for transferring groups other than "hydride". A first original application made use of biomimetic NAD+ models as "chiral amide-transferring agent" and assessed in "atropenantioselective amidification" of benzamides.

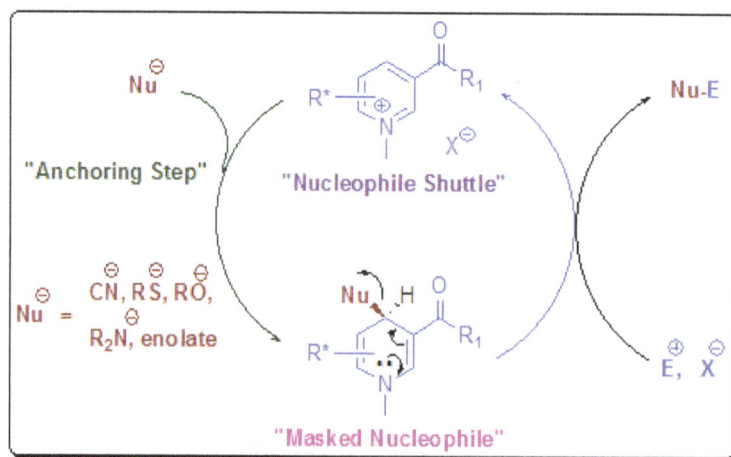

NAD$^+$ models as "chiral amide-transferring agent"

Unprecedented development of quinoliniun salts, structurally related to NAD+ models are currently exploited in peptide bond formation. In 2005, Vincent Levacher et al. demonstrated that quinolinium thioester salts-type 2 display an attractive potential in peptide synthesis. Interestingly, a number of experimental observations lend to the belief that a sequential mechanism related to a prior amine capture strategy is involved. The "latent reactivity" of the nonquaternized quinoline 1 renders this precursor an appealing synthetic tool in view of developing a new safety-catch linker. These preliminary results laid down the basis of future developments in SPPS.

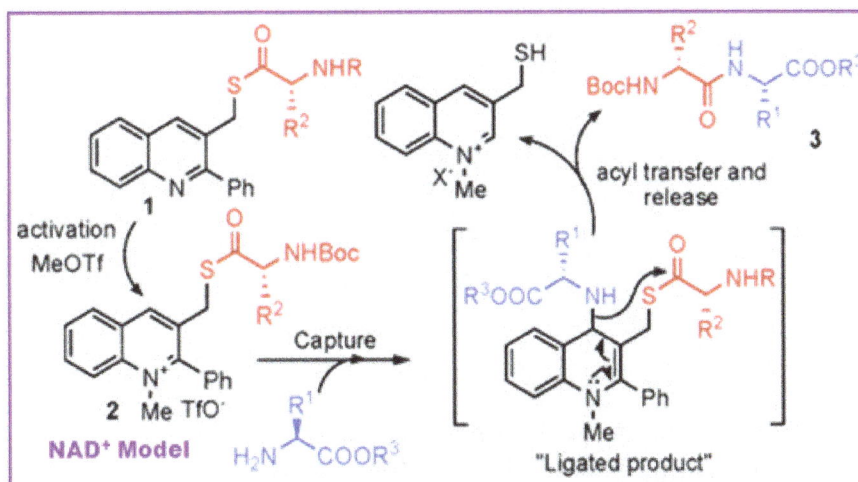

Amine capture strategy and peptide bond formation by means of a quinolinium thioester salt.

Structural principles and characterization (e.g. sugars: anomers of glucose, anomeric effect, diastereomers, NMR).

Application of biology to stereoselective chemical synthesis (e.g. yeast)

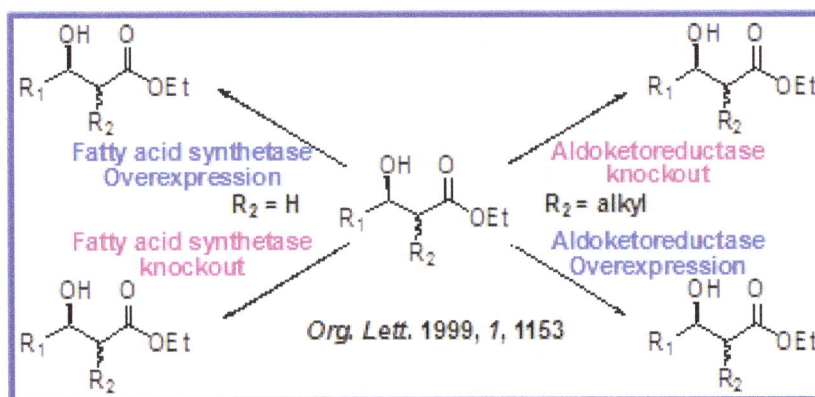

Synthesis of small molecules (e.g. peptides, drugs, dilantin, esters).

Chemical catalysis (e.g. protection & activation strategies relevant to peptide synthesis in vivo and in vitro).

Comparison of organic and biological reactions.

Enzyme mechanisms and active sites

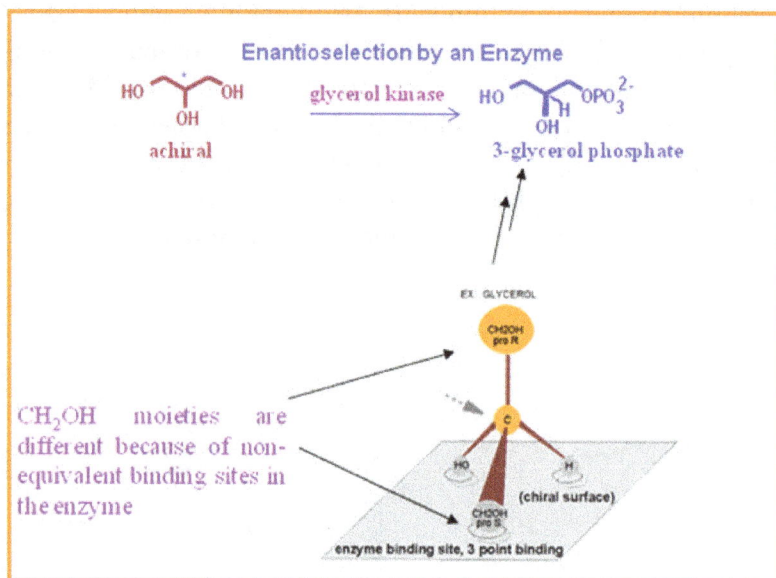

Example of Biochemical Knowledge Applied to Organic Chemistry:

1. Natural Product Chemistry: Synthesis of natural products like terpenoids, alkaloids can better be done from inquiry into the biosynthetic pathways of such natural products.

2. Biomimetic chemistry: It is a branch of organic chemistry wherein the object is to mimic natural reactions and enzymatic processes in order to get a better organic synthesis.

3. Pharmacology: designing drugs that inhibit a simple enzyme specifically, an example being the transition state analogs.

4. Enzyme Technology: A third example of biochemistry, applied to organic synthesis can be found in the growing field of enzyme technology.

Properties of Biological Molecules that Inspire Chemists:

1. Large → challenges: (a) for synthesis,

    (b) for structural prediction (e.g. protein folding)

2. Size → multiple Functional Groups (active site) aligned to achieve a goal (e.g. enzyme active site, bases in Nucleic Acids)

3. Multiple non-covalent weak interactions → strong, stable binding non-covalent complexes (e.g. substrate, inhibitor, DNA)

4. Specificity → specific interactions between 2 molecules in an ensemble within the cell

5. Regulated → switchable, allows control of cell → activation/inhibition

6. Catalysis → groups work in concert

7. Replication → turnover (e.g. an enzyme has many turnovers, nucleic acids replicate)

Example of Organic Chemical Knowledge for Understanding the Chemical Aspects of Life and its Origin:

Metabolism

It presents in all cells, which includes:-

1. Heredity: It means storage, transfer and expression of genetic information, with the underlying principle of paired organic bases, and

2. Storage, production and use of energy, with the possible underlying principle of paired moving charges.

Both processes together form metabolism, an intricate and strongly ordered system of chemical reactions, regulated by their catalysts, which are intricate and strongly ordered amino acid polymers. This central position of the enzymes in the living cell makes it understandable that research on the mechanisms of the enzymatic reactions presently takes a central place in this bioorganic chemistry.

Key Processes of Metabolism

1. Bases + sugars $\rightarrow$ nucleosides $\rightleftharpoons$ nucleic acids
2. Sugars (monosaccharides) $\rightleftharpoons$ polysaccharides
3. Amino acids $\rightleftharpoons$ proteins
4. Polymerization reactions: cell also needs the reverse process

If we want to look at each of these processes, forwards and backwards, by comparing and contrasting the reactions, we should raise questions:-

a. How do chemists synthesize these structures?

b. How might these structures have formed in the pre-biotic world, and have led to life on earth?

c. How are they made in vivo?

d. Can we design improved chemistry by understanding the biology: biomimetic synthesis?

It becomes steadily more customary to analyze possible mechanisms on the basis of fairly simple organic chemical models that combine only the more fundamental factors of the enzymatic catalysis. Starting from established reaction theories from both biochemistry and organic chemistry one tries to reconcile two ways of thinking in order to get new insight into what life is. The method itself, this roundabout way via relatively simple models, often leads to the surprising consequence that the problem of the origin of life emerges. The border between organic chemistry and biochemistry is man-made and systematic. This border is related closely to the historical transition between non-living and living. Some initial remarks concerning recent developments in the thinking about the origin of life, the "bioorganic era", therefore, are in order.

Evolution of Life

- Life did not suddenly crop up in its current form of complex structures (DNA, proteins) in one sudden reaction from mono-functional simple molecules

- We will follow some of the ideas of how life may have evolved:

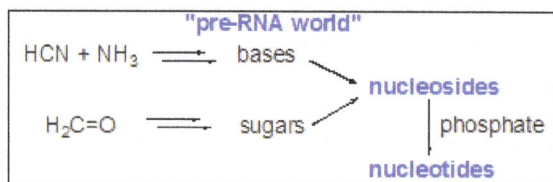

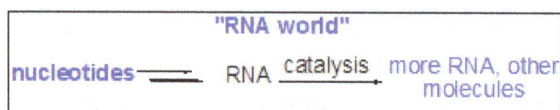

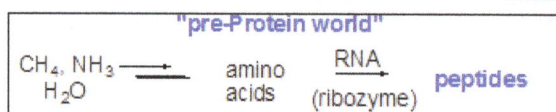

Molecular Recognition in Supramolecular Biochemistry

Proximity allows bond polarization which further results in an increase in the rate of the reaction of enzyme substrate interactions. It also reduces entropy. This process improves biological reactions like ligations or addition reactions. This chapter provides a plethora of interdisciplinary topics for better comprehension of proximity effect in organic chemistry.

Proximity Effect in Organic Chemistry

Proximity of reactive functional groups in a chemical transformation allows bond polarization, resulting generally in an acceleration of rate of the reaction.

Proximity and Orientation Effects

- This increases the rate of the reaction as enzyme-substrate interactions align reactive chemical groups and hold them close together. This reduces the entropy of the reactants and thus makes reactions such as ligations or addition reactions more favorable, there is a reduction in the overall loss of entropy when two reactants become a single product.
- This effect is analogous to an effective increase in concentration of the reagents. The binding of the reagents to the enzyme gives the reaction intramolecular character, which gives a massive rate increase.

Rate of a reaction depends on
- Number of collisions
- Energy of molecules
- Orientation of molecules
- Reaction pathway (transition state)

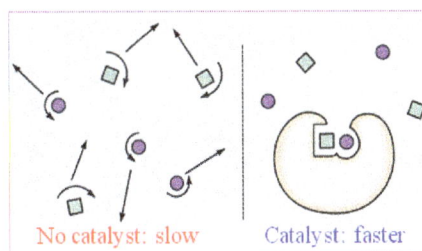

No catalyst: slow   Catalyst: faster

Proximity:
- Similar reactions will occur far faster if the reaction is intramolecular.
- Enezyme brings the two or more reactant closer in proximity to react.
- Brings substrate with their catalytic groups closer.
- Electrostatic catalysis:
  - transition state stabilization the charge distribution around the active sites guide polar substrates toward their binding site.

- Freezing out the molecular motions:transition state stabilization:

  - translational and rotational motions of their substrates and catalytic groupsrate enhancement up to ~10^7

Orientation:

- Enzymes not only bring substrates and catalytic groups close together, they orient them in a manner suitable for catalysis as well. Comparison of the rates of reaction of the molecules shown makes it clear that the bulky methyl groups force an orientation on the alkyl carboxylate and the aromatic hydroxyl groups that makes them approximately 250 billion times more likely to react. Enzymes function similarly by placing catalytically functional groups (from the protein side chains or from another substrate) in the proper position for reaction.

Some Example of Proximity Effects

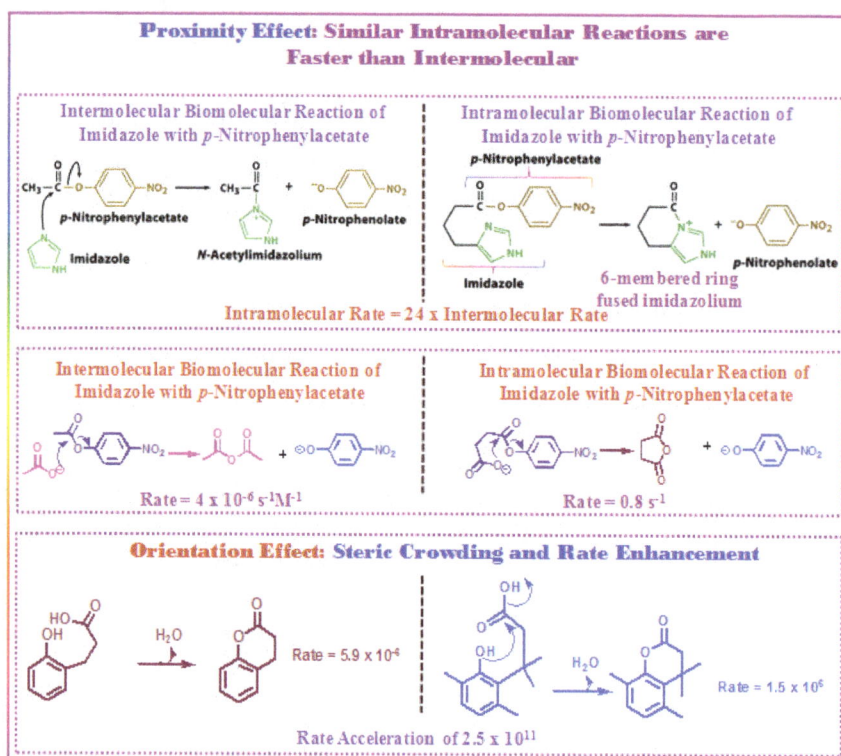

Examples depict proximity effects.

Molecular Recognition

Molecular recognition is the specific interaction between two or more molecules through no covalent bonding such as hydrogen bonding, metal coordination, hydrophobic forces, van deer Waals forces, pi-pi interactions, electrostatic and/or electromagnetic effects. The host and guest involved in molecular recognition exhibit molecular complementarities.

"Molecular recognition" covers a set of phenomena controlled by specific noncovalent interactions. Such phenomena are crucial in biological systems, and much modern chemical research. "Molecular recognition", which may be both inter- and intramolecular phenomena, is also encompasses the "host–guest chemistry", "supramolecular chemistry", and "self-assembly", though these are limited to intermolecular processes. Protein folding is a classic example of intramolecular recognition. It is the Host-Guest Interactions and in enzymology lock and key interaction.

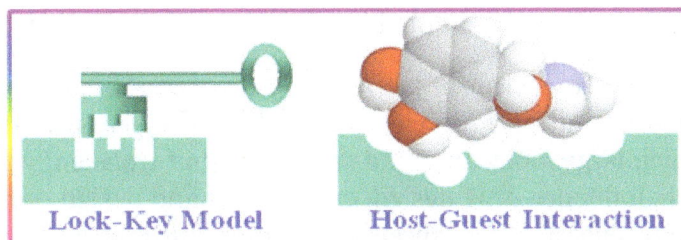

Schematic presentation of "Lock and key" interaction in enzymology and the "host–guest chemistry".

Types of Molecular Recognition: Static vs. Dynamic

Molecular recognition can be subdivided into static molecular recognition and dynamic molecular recognition. Static molecular recognition is likened to the interaction between a key and a keyhole; it is a 1:1 type complexation reaction between a host molecule and a guest molecule to form a host-guest complex. To achieve advanced static molecular recognition, it is necessary to make recognition sites that are specific for guest molecules.

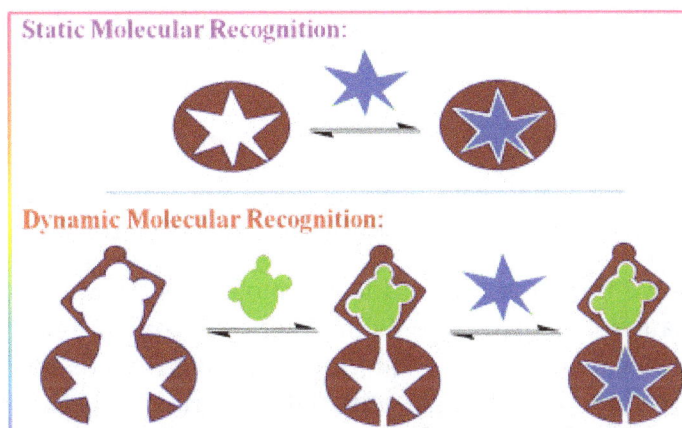

Static recognition between a single guest and a single host binding site. In dynamic recognition binding the first guest at the first binding site induces a conformation change that affects the association constant of the second guest at the second binding site, a positive allosteric site.

In the case of dynamic molecular recognition the binding of the first guest to the first binding site of a host affects the association constant of a second guest with a second binding site. In the case of positive allosteric systems the binding of the first guest increases the association constant of the second guest. While for negative allosteric systems the binding of the first guest decreases the association constant with the second. The dynamic nature of this type of molecular recognition is particularly important since it provides a mechanism to regulate binding in biological systems. Dynamic molecular recognition is also being studied for application in highly functional chemical sensors and molecular devices.

Non-covalent Interactions

A non-covalent interaction differs from a covalent bond in that it does not involve the sharing of electrons, but rather involves more dispersed variations of electromagnetic interactions between molecules or within a molecule. The chemical energy released in the formation of non-covalent interactions is typically on the order of 1-5 kcal/mol (1000–5000 calories per 6.02×10^{23} molecules). Non-covalent interactions can be classified into different categories, such as electrostatic, π-effects, van der Waals forces, and hydrophobic effects.

Non-covalent interactions are critical in maintaining the three-dimensional structure of large molecules, such as proteins and nucleic acids. In addition, they are also involved in many biological processes in which large molecules bind specifically but transiently to one another. These interactions also heavily influence drug design, crystallinity and design of materials, particularly for self-assembly, and, in general, the synthesis of many organic molecules.

Intermolecular forces are non-covalent interactions that occur between different molecules, rather than between different atoms of the same molecule.

Electrostatic Interactions

Ionic

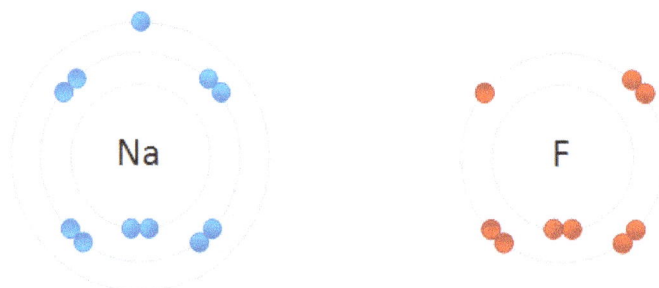

Process of NaF formation -- example of an electrostatic interaction

Ionic interactions involve the attraction of ions or molecules with full permanent charges of opposite signs. For example, sodium fluoride involves the attraction of the positive charge on sodium (Na^+) with the negative charge on fluoride (F^-). These bonds are harder to break than covalent bonds because there is a strong electrostatic interaction between oppositely charged ions. However, this particular interaction is easily broken upon addition to water, or other highly polar solvents.

These interactions can also be seen in molecules with a localized charge on a particular atom. For example, the full negative charge associated with ethoxide, the conjugate base of ethanol, is most commonly accompanied by the positive charge of an alkali metal salt such as the sodium cation (Na^+).

Hydrogen Bonding

A hydrogen bond (H-bond), is a specific type of interaction that involves dipole-dipole attraction between a partially positive hydrogen atom and a highly electronegative, partially negative oxygen, nitrogen, sulfur, or fluorine atom (not covalently bound to said hydrogen atom). It is not a covalent bond, but instead is classified as a strong non-covalent interaction. It is responsible for why water is a liquid at room temperature and not a gas (given water's low molecular weight). Most commonly, the strength of hydrogen bonds lies between 0 - 4 kcal/mol, but can sometimes be as strong as 40 kcal/mol.

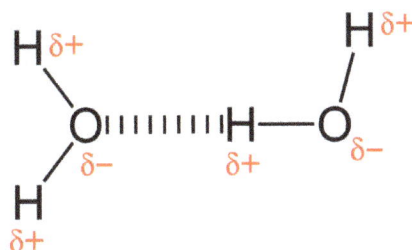

Halogen Bonding

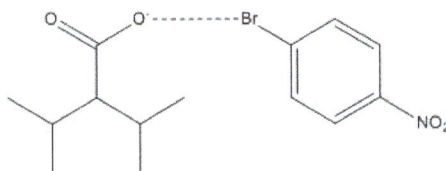

Anionic Lewis Base forming a halogen bond with electron-withdrawn bromine (Lewis acid)

Halogen bonding is a type of non-covalent interaction which does not involve the formation nor breaking of actual bonds, but rather is similar to the dipole-dipole interaction known as hydrogen bonding. In halogen bonding, a halogen atom acts as an electrophile, or electron-seeking species, and forms a weak electrostatic interaction with a nucleophile, or electron-rich species. The nucleophilic agent in these interactions tends to be highly electronegative (such as oxygen, nitrogen, or sulfur), or may be anionic, bearing a negative formal charge. As compared to hydrogen bonding, the halogen atom takes the place of the partially positively charged hydrogen as the electrophile.

Halogen bonding should not be confused with halogen-aromatic interactions, as the two are related but differ by definition. Halogen-aromatic interactions involve an electron-rich aromatic π-cloud as a nucleophile; halogen bonding is restricted to monatomic nucleophiles.

Van der Waals Forces

Van der Waals Forces are a subset of electrostatic interactions involving permanent or induced dipoles (or multipoles). These include the following:

- permanent dipole-dipole interactions, alternatively called the Keesom force

- dipole-induced dipole interactions, or the Debye force

- induced dipole-induced dipole interactions, commonly referred to as London dispersion forces

Hydrogen bonding and halogen bonding are typically not classified as Van der Waals forces.

Dipole-dipole

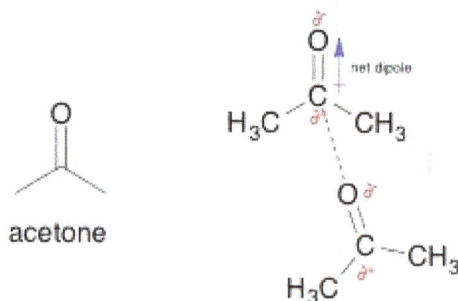

Dipole-dipole interactions between two acetone molecules,
with the partially negative oxygen atom interacting with the partially positive carbon atom in the carbonyl.

Dipole-dipole interactions are electrostatic interactions between permanent dipoles in molecules. These interactions tend to align the molecules to increase attraction (reducing potential energy). Normally, dipoles are associated with electronegative atoms, including oxygen, nitrogen, sulfur, and fluorine.

For example, acetone, the active ingredient in some nail polish removers, has a net dipole associated with the carbonyl. Since oxygen is more electronegative than the carbon that is covalently bonded to it, the electrons associated with that bond will be closer to the oxygen than the carbon, creating a partial negative charge (δ^-) on the oxygen, and a partial positive charge (δ^+) on the carbon. They are not full charges because the electrons are still shared through a covalent bond between the oxygen and carbon. If the electrons were no longer being shared, then the oxygen-carbon bond would be an electrostatic interaction.

$$\overset{\delta+}{H}-\overset{\delta-}{Cl}\cdots\overset{\delta+}{H}-\overset{\delta-}{Cl}$$

Often molecules contain dipolar groups, but have no overall dipole moment. This occurs if there is symmetry within the molecule that causes the dipoles to cancel each other out. This occurs in molecules such as tetrachloromethane. Note that the dipole-dipole interaction between two individual atoms is usually zero, since atoms rarely carry a permanent dipole.

Dipole-induced Dipole

A dipole-induced dipole interaction (Debye force) is due to the approach of a molecule with a permanent dipole to another non-polar molecule with no permanent dipole. This approach causes the electrons of the non-polar molecule to be polarized toward or away from the dipole (or "induce" a dipole) of the approaching molecule. Specifically, the dipole can cause electrostatic attraction or repulsion of the electrons from the non-polar molecule, depending on orientation of the incoming dipole. Atoms with larger atomic radii are considered more "polarizable" and therefore experience greater attraction as a result of the Debye force.

London Dispersion Forces

London dispersion forces are the weakest type of non-covalent interaction. They are also known as "induced dipole-induced dipole interactions" and present between all molecules, even those which inherently do not have permanent dipoles. They are caused by the temporary repulsion of electrons away from the electrons of a neighboring molecule, leading to a partially positive dipole on one molecule and a partially negative dipole on another molecule. Hexane is a good example of a molecule with no polarity or highly electronegative atoms, yet is a liquid at room temperature due mainly to London dispersion forces. In this example, when one hexane molecule approaches another, a temporary, weak partially negative dipole on the incoming hexane can polarize the electron cloud of another, causing a partially positive dipole on that hexane molecule. While these interactions are short-lived and very weak, they can be responsible for why certain non-polar molecules are liquids at room temperature.

π-effects

π-effects can be broken down into numerous categories, including π-π interactions, cation-π & anion-π interactions, and polar-π interactions. In general, π-effects are associated with the interactions of molecules with the π-systems of conjugated molecules such as benzene.

π-π Interaction

π-π interactions are associated with the interaction between the π-orbitals of a molecular system. For a simple example, a benzene ring, with its fully conjugated π cloud, will interact in two major ways (and one minor way) with a neighboring benzene ring through a π-π interaction. The two major ways that benzene stacks are edge-to-face, with an enthalpy of ~2 kcal/mol, and displaced (or slip stacked), with an enthalpy of ~2.3 kcal/mol. Interestingly, the sandwich configuration is not nearly as stable of an interaction as the previously two mentioned due to high electrostatic repulsion of the electrons in the π orbitals.

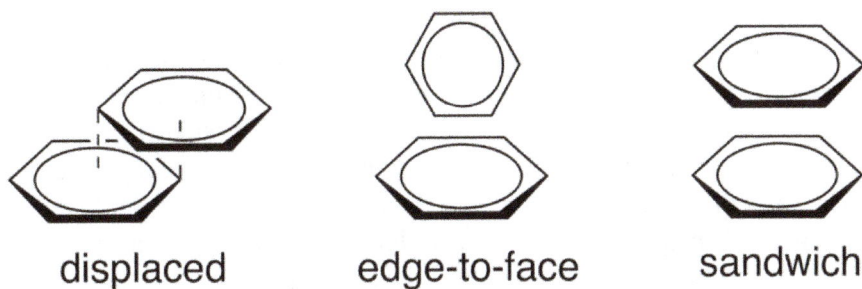

displaced   edge-to-face   sandwich

Various ways that benzene can interact intermolecularly. Note, however,
that the sandwich configuration is not a favorable interaction compared to displaced or edge-to-face

Cation-π and Anion-π Interaction

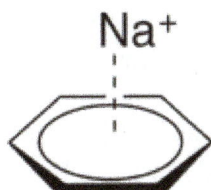

Na+

Cation-π interactions involve the positive charge of a cation interacting with the electrons in a π-system of a molecule. This interaction is surprisingly strong (as strong or stronger than H-bonding in some contexts), and has many potential applications in chemical sensors. For example, the sodium ion can easily sit atop the π cloud of a benzene molecule, with C_6 symmetry.

Anion-π interactions are very similar to cation-π interactions, but reversed. In this case, an anion sits atop an electron-poor π-system, usually established by the placement of electron-withdrawing substituents on the conjugated molecule

Polar-π

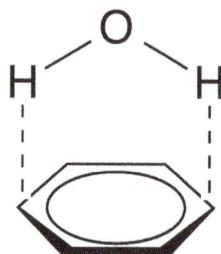

Polar-π interactions involve molecules with permanent dipoles (such as water) interacting with the quadrupole moment of a π-system (such as that in benzene. While not as strong as a cation-π interaction, these interactions can be quite strong (~1-2 kcal/mol), and are commonly involved in protein folding and crystallinity of solids containing both hydrogen bonding and π-systems. In fact, any molecule with a hydrogen bond donor (hydrogen bound to a highly electronegative atom) will have favorable electrostatic interactions with the electron-rich π-system of a conjugated molecule.

Hydrophobic Effect

The hydrophobic effect is the desire for non-polar molecules to aggregate in aqueous solutions in order to separate from water. This phenomenon leads to minimum exposed surface area of non-polar molecules to the polar water molecules (typically spherical droplets), and is commonly used in biochemistry to study protein folding and other various biological phenomenon. The effect is also commonly seen when mixing various oils (including cooking oil) and water. Over time, oil sitting on top of water will begin to aggregate into large flattened spheres from smaller droplets, eventually leading to a film of all oil sitting atop a pool of water.

Examples

Drug Design

Most pharmaceutical drugs are small molecules which elicit a physiological response by "binding" to enzymes or receptors, causing an increase or decrease in the enzyme's ability to function. The binding of a small molecule to a protein is governed by a combination of steric, or spatial considerations, in addition to various non-covalent interactions, although some drugs do covalently modify an active site. Using the "lock and key model" of enzyme binding, a drug (key) must be of roughly the proper dimensions to fit the enzyme's binding site (lock). Using the appropriately sized molecular scaffold, drugs must also interact with the enzyme non-covalently in order to maximize binding affinity binding constant and reduce the ability of the drug to dissociate from the binding site. This is achieved by forming various non-covalent interactions between the small molecule and amino acids in the binding site, including: hydrogen bonding, electrostatic interactions, pi stacking, van der Waals interactions, and dipole-dipole interactions.

Protein Folding and Structure

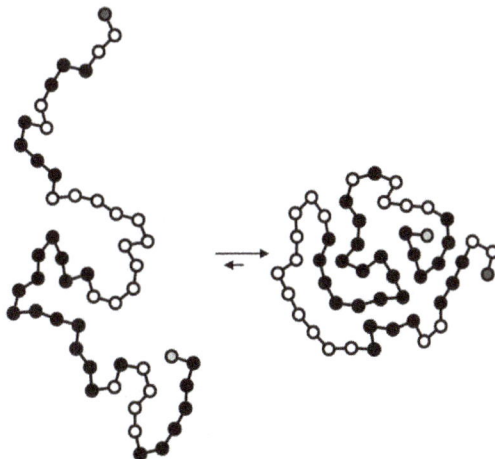

Illustration of the main driving force behind protein structure formation. In the compact fold (to the right), the hydrophobic amino acids (shown as black spheres) are in general shielded from the solvent.

The folding of most proteins from a primary (linear) sequence of amino acids to a three-dimensional structure is governed by many factors, including non-covalent interactions. The first ~5 milliseconds of folding are primarily dependent on van der Waals forces, whereby the protein folds so as to orient nonpolar amino acids in the interior of the globular protein, while more polar amino acid residues are exposed to aqueous solvent. This phase is known as the hydrophobic collapse, when nonpolar non-covalent interactions exclude water from the interior of the developing 3D protein structure.

After this initial "burst phase," more polar non-covalent interactions take over. Between 5 and 1000 milliseconds after protein folding initiation, three-dimensional structures of proteins, known as secondary and tertiary structures, are stabilized by formation of hydrogen bonds, in addition to disulfide bridges (covalent linkages). Through a series of small conformational changes, spatial orientations are modified so as to arrive at the most energetically minimized orientation achievable. The folding of proteins is often facilitated by enzymes known as molecular chaperones Sterics, bond strain, and angle strain also play major roles in the folding of a protein from its primary sequence to its tertiary structure.

Single tertiary protein structures can also assemble to form protein complexes composed of multiple independently folded subunits. As a whole, this is called a protein's quaternary structure. The quaternary structure is generated by the formation of relatively strong non-covalent interactions, such as hydrogen bonds, between different subunits to generate a functional polymeric enzyme. Some proteins also utilize non-covalent interactions to bind cofactors in the active site during catalysis, however a cofactor can also be covalently attached to an enzyme. Cofactors can be either organic or inorganic molecules which assist in the catalytic mechanism of the active enzyme. The strength with which a cofactor is bound to an enzyme may vary greatly; non-covalently bound cofactors are typically anchored by hydrogen bonds or electrostatic interactions.

Boiling Points

Non-covalent interactions have a significant effect on the boiling point of a liquid. Boiling point is defined as the temperature at which the vapor pressure of a liquid is equal to the pressure surrounding the liquid. More simply, it is the temperature at which a liquid becomes a gas. As one might expect, the stronger the non-covalent interactions present for a substance, the higher its boiling point. For example, consider three compounds of similar chemical composition: sodium n-butoxide (C_4H_9ONa), diethyl ether ($C_4H_{10}O$), and n-butanol (C_4H_9OH).

Name	Sodium n-butoxide	n-butanol	Diethyl ether
Chemical Formula	$C_4H_7O^-$ Na^+	C_4H_7OH	$C_4H_{10}O$
Strongest non-covalent interaction	Ionic interaction	Hydrogen bonding	Dipole-dipole
Boiling Point	>260 °C	117 °C	35 °C

Boiling points of 4-carbon compounds

The predominant non-covalent interactions associated with each species in solution are listed in the above figure. As previously discussed, ionic interactions require considerably more energy to break than hydrogen bonds, which in turn are require more energy than dipole-dipole interactions. The trends observed in their boiling points shows exactly the correlation expected, where sodium n-butoxide requires significantly more heat energy (higher temperature) to boil than n-butanol, which boils at a much higher temperature than diethyl ether. The heat energy required for a compound to change from liquid to gas is associated with the energy required to break the intermolecular forces each molecule experiences in its liquid state.

Hydrogen Bond

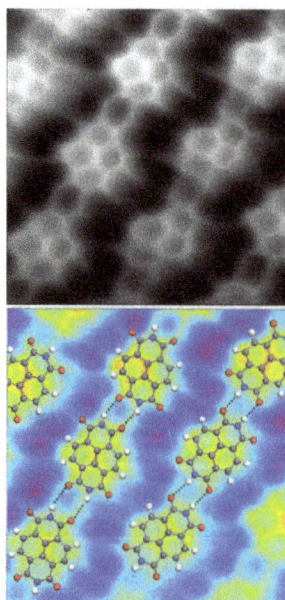

AFM image of napthalenetetracarboxylic diimide molecules on silver-terminated silicon, interacting via hydrogen bonding, taken at 77 K. ("Hydrogen bonds" in the top image are exaggerated by artifacts of the imaging technique.)

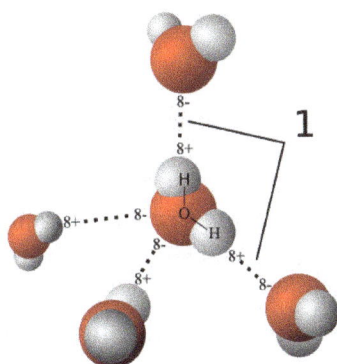

Model of hydrogen bonds (1) between molecules of water

A hydrogen bond is the electrostatic attraction between two polar groups that occurs when a hydrogen (H) atom covalently bound to a highly electronegative atom such as nitrogen (N), oxygen (O), or fluorine (F) experiences the electrostatic field of another highly electronegative atom nearby.

Hydrogen bonds can occur between molecules (*intermolecular*) or within different parts of a single molecule (*intramolecular*). Depending on geometry and environment, the hydrogen bond free energy content is between 1 and 5 kcal/mol. This makes it stronger than a van der Waals interaction, but weaker than covalent or ionic bonds. This type of bond can occur in inorganic molecules such as water and in organic molecules like DNA and proteins.

Intermolecular hydrogen bonding is responsible for the high boiling point of water (100°C) compared to the other group 16 hydrides that have much weaker hydrogen bonds. Intramolecular hydrogen bonding is partly responsible for the secondary and tertiary structures of proteins and nucleic acids. It also plays an important role in the structure of polymers, both synthetic and natural.

The hydrogen bond is an attractive interaction between a hydrogen atom from a molecule or a molecular fragment X–H in which X is more electronegative than H, and an atom or a group of atoms in the same or a different molecule, in which there is evidence of bond formation.

An accompanying detailed technical report provides the rationale behind the new definition.

Bonding

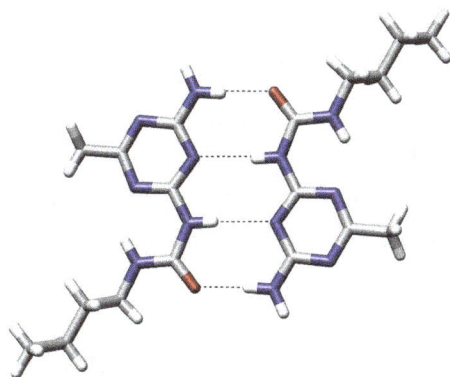

An example of intermolecular hydrogen bonding in a self-assembled dimer complex reported by Meijer and coworkers. The hydrogen bonds are represented by dotted lines.

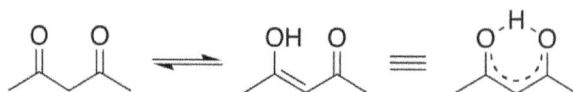

Intramolecular hydrogen bonding in acetylacetone helps stabilize the enol tautomer.

A hydrogen atom attached to a relatively electronegative atom will play the role of the hydrogen bond *donor*. This electronegative atom is usually fluorine, oxygen, or nitrogen. A hydrogen attached to carbon can also participate in hydrogen bonding when the carbon atom is bound to electronegative atoms, as is the case in chloroform, $CHCl_3$. An example of a hydrogen bond donor is the hydrogen from the hydroxyl group of ethanol, which is bonded to an oxygen.

In a hydrogen bond, the electronegative atom not covalently attached to the hydrogen is named proton acceptor, whereas the one covalently bound to the hydrogen is named the proton donor.

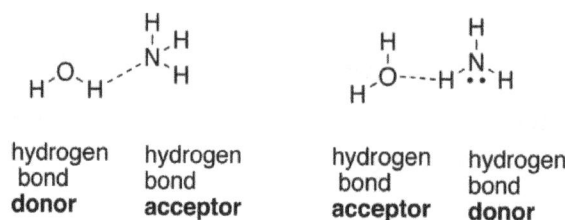

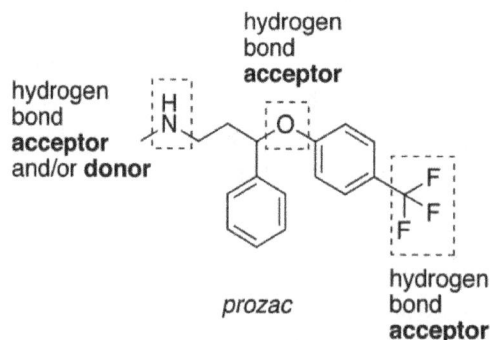

Examples of hydrogen bond donating (donors) and hydrogen bond accepting groups (acceptors)

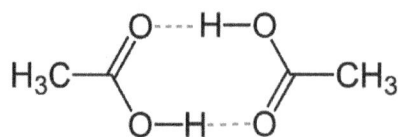

Cyclic dimer of acetic acid; dashed **green** lines represent hydrogen bonds

In the donor molecule, the electronegative atom attracts the electron cloud from around the hydrogen nucleus of the donor, and, by decentralizing the cloud, leaves the atom with a positive partial charge. Because of the small size of hydrogen relative to other atoms and molecules, the resulting charge, though only partial, represents a large charge density. A hydrogen bond results when this strong positive charge density attracts a lone pair of electrons on another heteroatom, which then becomes the hydrogen-bond acceptor.

The hydrogen bond is often described as an electrostatic dipole-dipole interaction. However, it also has some features of covalent bonding: it is directional and strong, produces interatomic distances shorter than the sum of the van der Waals radii, and usually involves a limited number of interaction partners, which can be interpreted as a type of valence. These covalent features are more substantial when acceptors bind hydrogens from more electronegative donors.

The partially covalent nature of a hydrogen bond raises the following questions: "To which molecule or atom does the hydrogen nucleus belong?" and "Which should be labeled 'donor' and which 'acceptor'?" Usually, this is simple to determine on the basis of interatomic distances in the X–H⋯Y system, where the dots represent the hydrogen bond: the X–H distance is typically ≈110 pm, whereas the H⋯Y distance is ≈160 to 200 pm. Liquids that display hydrogen bonding (such as water) are called associated liquids.

Hydrogen bonds can vary in strength from very weak (1–2 kJ mol^{-1}) to extremely strong (161.5 kJ mol^{-1} in the ion HF_2^-). Typical enthalpies in vapor include:

- F–H⋯:F (161.5 kJ/mol or 38.6 kcal/mol)

- O–H⋯:N (29 kJ/mol or 6.9 kcal/mol)

- O–H⋯:O (21 kJ/mol or 5.0 kcal/mol)

- N–H⋯:N (13 kJ/mol or 3.1 kcal/mol)

- N–H⋯:O (8 kJ/mol or 1.9 kcal/mol)

- HO–H⋯:OH_3^+(18 kJ/mol or 4.3 kcal/mol; data obtained using molecular dynamics as detailed in the reference and should be compared to 7.9 kJ/mol for bulk water, obtained using the same molecular dynamics.)

Quantum chemical calculations of the relevant interresidue potential constants (compliance constants) revealed large differences between individual H bonds of the same type. For example, the central interresidue N–H⋯N hydrogen bond between guanine and cytosine is much stronger in comparison to the N–H⋯N bond between the adenine-thymine pair.

The length of hydrogen bonds depends on bond strength, temperature, and pressure. The bond strength itself is dependent on temperature, pressure, bond angle, and environment (usually characterized by local dielectric constant). The typical length of a hydrogen bond in water is 197 pm. The ideal bond angle depends on the nature of the hydrogen bond donor. The following hydrogen bond angles between a hydrofluoric acid donor and various acceptors have been determined experimentally:

Acceptor⋯donor	VSEPR geometry	Angle (°)
HCN⋯HF	linear	180
H_2CO⋯HF	trigonal planar	120
H_2O⋯HF	pyramidal	46
H_2S⋯HF	pyramidal	89
SO_2⋯HF	trigonal	142

History

In the book *The Nature of the Chemical Bond*, Linus Pauling credits T. S. Moore and T. F. Winmill with the first mention of the hydrogen bond, in 1912. Moore and Winmill used the hydrogen bond to account for the fact that trimethylammonium hydroxide is a weaker base than tetramethylammonium hydroxide. The description of hydrogen bonding in its better-known setting, water, came some years later, in 1920, from Latimer and Rodebush. In that paper, Latimer and Rodebush cite work by a fellow scientist at their laboratory, Maurice Loyal Huggins, saying, "Mr. Huggins of this laboratory in some work as yet unpublished, has used the idea of a hydrogen kernel held between two atoms as a theory in regard to certain organic compounds."

Hydrogen Bonds in Water

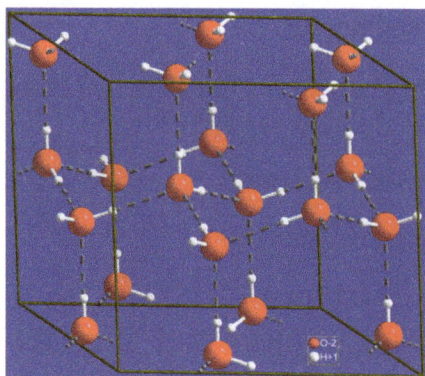

Crystal structure of hexagonal ice. Gray dashed lines indicate hydrogen bonds

The most ubiquitous and perhaps simplest example of a hydrogen bond is found between water molecules. In a discrete water molecule, there are two hydrogen atoms and one oxygen atom. Two molecules of water can form a hydrogen bond between them; the simplest case, when only two molecules are present, is called the water dimer and is often used as a model system. When more molecules are present, as is the case with liquid water, more bonds are possible because the oxygen of one water molecule has two lone pairs of electrons, each of which can form a hydrogen bond with a hydrogen on another water molecule. This can repeat such that every water molecule is H-bonded with up to four other molecules, as shown in the figure (two through its two lone pairs, and two through its two hydrogen atoms). Hydrogen bonding strongly affects the crystal structure of ice, helping to create an open hexagonal lattice. The density of ice is less than the density of water at the same temperature; thus, the solid phase of water floats on the liquid, unlike most other substances.

Liquid water's high boiling point is due to the high number of hydrogen bonds each molecule can form, relative to its low molecular mass. Owing to the difficulty of breaking these bonds, water has a very high boiling point, melting point, and viscosity compared to otherwise similar liquids not conjoined by hydrogen bonds. Water is unique because its oxygen atom has two lone pairs and two hydrogen atoms, meaning that the total number of bonds of a water molecule is up to four. For example, hydrogen fluoride—which has three lone pairs on the F atom but only one H atom—can form only two bonds; (ammonia has the opposite problem: three hydrogen atoms but only one lone pair).

$$H-F\cdots H-F\cdots H-F$$

The exact number of hydrogen bonds formed by a molecule of liquid water fluctuates with time and depends on the temperature. From TIP4P liquid water simulations at 25 °C, it was estimated that each water molecule participates in an average of 3.59 hydrogen bonds. At 100 °C, this number decreases to 3.24 due to the increased molecular motion and decreased density, while at 0 °C, the average number of hydrogen bonds increases to 3.69. A more recent study found a much smaller number of hydrogen bonds: 2.357 at 25 °C. The differences may be due to the use of a different method for defining and counting the hydrogen bonds.

Where the bond strengths are more equivalent, one might instead find the atoms of two interacting water molecules partitioned into two polyatomic ions of opposite charge, specifically hydroxide (OH^-) and hydronium (H_3O^+). (Hydronium ions are also known as "hydroxonium" ions.)

$$H-O^-\ H_3O^+$$

Indeed, in pure water under conditions of standard temperature and pressure, this latter formulation is applicable only rarely; on average about one in every 5.5×10^8 molecules gives up a proton to another water molecule, in accordance with the value of the dissociation constant for water under such conditions. It is a crucial part of the uniqueness of water.

Because water may form hydrogen bonds with solute proton donors and acceptors, it may competitively inhibit the formation of solute intermolecular or intramolecular hydrogen bonds. Consequently, hydrogen bonds between or within solute molecules dissolved in water are almost always unfavorable relative to hydrogen bonds between water and the donors and acceptors for hydrogen bonds on those solutes. Hydrogen bonds between water molecules have an average lifetime of 10^{-11} seconds, or 10 picoseconds.

Bifurcated and Over-coordinated Hydrogen Bonds in Water

A single hydrogen atom can participate in two hydrogen bonds, rather than one. This type of bonding is called "bifurcated" (split in two or "two-forked"). It can exist, for instance, in complex natural or synthetic organic molecules. It has been suggested that a bifurcated hydrogen atom is an essential step in water reorientation.

Acceptor-type hydrogen bonds (terminating on an oxygen's lone pairs) are more likely to form bifurcation (it is called overcoordinated oxygen, OCO) than are donor-type hydrogen bonds, beginning on the same oxygen's hydrogens.

Hydrogen Bonds in DNA and Proteins

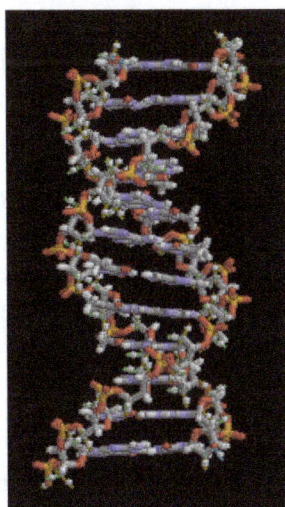

The structure of part of a DNA double helix

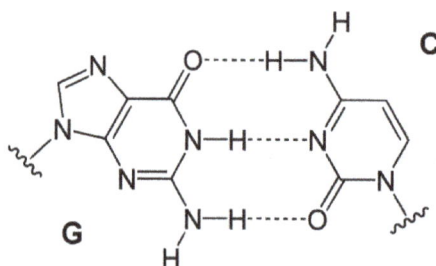

Hydrogen bonding between guanine and cytosine, one of two types of base pairs in DNA.

Hydrogen bonding also plays an important role in determining the three-dimensional structures adopted by proteins and nucleic bases. In these macromolecules, bonding between parts of the same macromolecule cause it to fold into a specific shape, which helps determine the molecule's physiological or biochemical role. For example, the double helical structure of DNA is due largely to hydrogen bonding between its base pairs (as well as pi stacking interactions), which link one complementary strand to the other and enable replication.

In the secondary structure of proteins, hydrogen bonds form between the backbone oxygens and amide hydrogens. When the spacing of the amino acid residues participating in a hydrogen bond occurs regularly between positions i and $i + 4$, an alpha helix is formed. When the spacing is less, between positions i and $i + 3$, then a 3_{10} helix is formed. When two strands are joined by hydrogen bonds involving alternating residues on each participating strand, a beta sheet is formed. Hydrogen bonds also play a part in forming the tertiary structure of protein through interaction of R-groups..

The role of hydrogen bonds in protein folding has also been linked to osmolyte-induced protein stabilization. Protective osmolytes, such as trehalose and sorbitol, shift the protein folding equilibrium toward the folded state, in a concentration dependent manner. While the prevalent explanation for osmolyte action relies on excluded volume effects, that are entropic in nature, recent Circular dichroism (CD) experiments have shown osmolyte to act through an enthalpic effect. The molecular mechanism for their role in protein stabilization is still not well established, though several mechanism have been proposed. Recently, computer molecular dynamics simulations suggested that osmolytes stabilize proteins by modifying the hydrogen bonds in the protein hydration layer.

Several studies have shown that hydrogen bonds play an important role for the stability between subunits in multimeric proteins. For example, a study of sorbitol dehydrogenase displayed an important hydrogen bonding network which stabilizes the tetrameric quaternary structure within the mammalian sorbitol dehydrogenase protein family.

A protein backbone hydrogen bond incompletely shielded from water attack is a dehydron. Dehydrons promote the removal of water through proteins or ligand binding. The exogenous dehydration enhances the electrostatic interaction between the amide and carbonyl groups by de-shielding their partial charges. Furthermore, the dehydration stabilizes the hydrogen bond by destabilizing the nonbonded state consisting of dehydrated isolated charges.

Hydrogen Bonds in Polymers

Many polymers are strengthened by hydrogen bonds in their main chains. Among the synthetic polymers, the best known example is nylon, where hydrogen bonds occur in the repeat unit and play a major role in crystallization of the material. The bonds occur between carbonyl and amine groups in the amide repeat unit. They effectively link adjacent chains to create crystals, which help reinforce the material. The effect is greatest in aramid fibre, where hydrogen bonds stabilize the linear chains laterally. The chain axes are aligned along the fibre axis, making the fibres extremely stiff and strong. Hydrogen bonds are also important in the structure of cellulose and derived polymers in its many different forms in nature, such as wood and natural fibres such as cotton and flax.

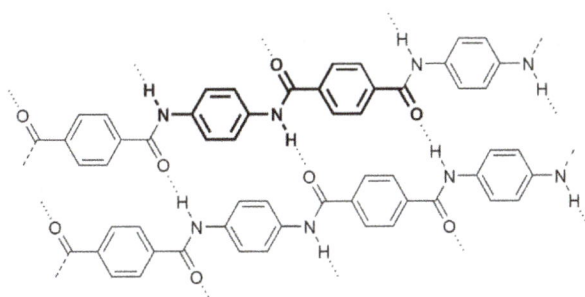

Para-aramid structure

The hydrogen bond networks make both natural and synthetic polymers sensitive to humidity levels in the atmosphere because water molecules can diffuse into the surface and disrupt the network. Some polymers are more sensitive than others. Thus nylons are more sensitive than aramids, and nylon 6 more sensitive than nylon-11.

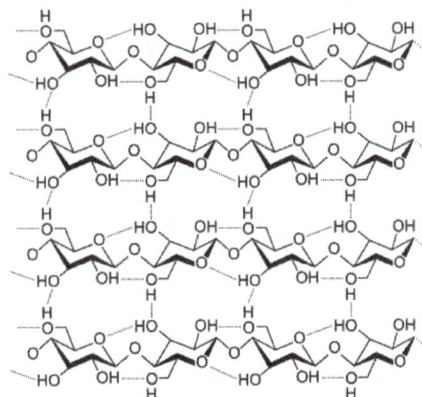

A strand of cellulose (conformation I_α), showing the hydrogen bonds (dashed) within and between cellulose molecules.

Symmetric Hydrogen Bond

A symmetric hydrogen bond is a special type of hydrogen bond in which the proton is spaced exactly halfway between two identical atoms. The strength of the bond to each of those atoms is equal. It is an example of a three-center four-electron bond. This type of bond is much stronger than a "normal" hydrogen bond. The effective bond order is 0.5, so its strength is comparable to a covalent bond. It is seen in ice at high pressure, and also in the solid phase of many anhydrous acids such as hydrofluoric acid and formic acid at high pressure. It is also seen in the bifluoride ion $[F-H-F]^-$.

Symmetric hydrogen bonds have been observed recently spectroscopically in formic acid at high pressure (>GPa). Each hydrogen atom forms a partial covalent bond with two atoms rather than one. Symmetric hydrogen bonds have been postulated in ice at high pressure (Ice X). Low-barrier hydrogen bonds form when the distance between two heteroatoms is very small.

Dihydrogen Bond

The hydrogen bond can be compared with the closely related dihydrogen bond, which is also an intermolecular bonding interaction involving hydrogen atoms. These structures have been known for some time, and well characterized by crystallography; however, an understanding of their relationship to the conventional hydrogen bond, ionic bond, and covalent bond remains unclear. Generally, the hydrogen bond is characterized by a proton acceptor that is a lone pair of electrons in nonmetallic atoms (most notably in the nitrogen, and chalcogen groups). In some cases, these proton acceptors may be pi-bonds or metal complexes. In the dihydrogen bond, however, a metal hydride serves as a proton acceptor, thus forming a hydrogen-hydrogen interaction. Neutron diffraction has shown that the molecular geometry of these complexes is similar to hydrogen bonds, in that the bond length is very adaptable to the metal complex/hydrogen donor system.

Advanced Theory of the Hydrogen Bond

In 1999, Isaacs *et al.* showed from interpretations of the anisotropies in the Compton profile of ordinary ice that the hydrogen bond is partly covalent. However, this interpretation was challenged by Ghanty et al., who concluded that considering electrostatic forces alone could explain the experimental results. Some NMR data on hydrogen bonds in proteins also indicate covalent bonding.

Most generally, the hydrogen bond can be viewed as a metric-dependent electrostatic scalar field between two or more intermolecular bonds. This is slightly different from the intramolecular bound states of, for example, covalent or ionic bonds; however, hydrogen bonding is generally still a bound state phenomenon, since the interaction energy has a net negative sum. The initial theory of hydrogen bonding proposed by Linus Pauling suggested that the hydrogen bonds had a partial covalent nature. This remained a controversial conclusion until the late 1990s when NMR techniques were employed by F. Cordier *et al.* to transfer information between hydrogen-bonded nuclei, a feat that would only be possible if the hydrogen bond contained some covalent character. While much experimental data has been recovered for hydrogen bonds in water, for example, that provide good resolution on the scale of intermolecular distances and molecular thermodynamics, the kinetic and dynamical properties of the hydrogen bond in dynamic systems remain unchanged.

Dynamics Probed by Spectroscopic Means

The dynamics of hydrogen bond structures in water can be probed by the IR spectrum of OH stretching vibration. In the hydrogen bonding network in protic organic ionic plastic crystals (POIPCs), which are a type of phase change material exhibiting solid-solid phase transitions prior to melting, variable-temperature infrared spectroscopy can reveal the temperature dependence of hydrogen bonds and the dynamics of both the anions and the cations. The sudden weakening of hydrogen bonds during the solid-solid phase transition seems to be coupled with the onset of orientational or rotational disorder of the ions.

Hydrogen Bonding Phenomena

- Dramatically higher boiling points of NH_3, H_2O, and HF compared to the heavier analogues PH_3, H_2S, and HCl.

- Increase in the melting point, boiling point, solubility, and viscosity of many compounds can be explained by the concept of hydrogen bonding.

- Occurrence of proton tunneling during DNA replication is believed to be responsible for cell mutations.

- Viscosity of anhydrous phosphoric acid and of glycerol

- Dimer formation in carboxylic acids and hexamer formation in hydrogen fluoride, which occur even in the gas phase, resulting in gross deviations from the ideal gas law.

- Pentamer formation of water and alcohols in apolar solvents.

- High water solubility of many compounds such as ammonia is explained by hydrogen bonding with water molecules.

- Negative azeotropy of mixtures of HF and water

- Deliquescence of NaOH is caused in part by reaction of OH^- with moisture to form hydrogen-bonded $H_3O_2^-$ species. An analogous process happens between $NaNH_2$ and NH_3, and between NaF and HF.

- The fact that ice is less dense than liquid water is due to a crystal structure stabilized by hydrogen bonds.

- The presence of hydrogen bonds can cause an anomaly in the normal succession of states of matter for certain mixtures of chemical compounds as temperature increases or decreases. These compounds can be liquid until a certain temperature, then solid even as the temperature increases, and finally liquid again as the temperature rises over the "anomaly interval"

- Smart rubber utilizes hydrogen bonding as its sole means of bonding, so that it can "heal" when torn, because hydrogen bonding can occur on the fly between two surfaces of the same polymer.

- Strength of nylon and cellulose fibres.

- Wool, being a protein fibre, is held together by hydrogen bonds, causing wool to recoil when stretched. However, washing at high temperatures can permanently break the hydrogen bonds and a garment may permanently lose its shape.

Coordination Complex

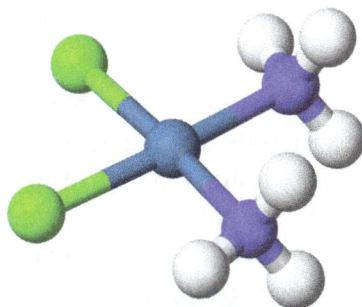

Cisplatin, $PtCl_2(NH_3)_2$, is a coordination complex of platinum(II) with two chloride and two ammonia ligands. It is one of the most successful anticancer drugs.

In chemistry, a coordination complex consists of a central atom or ion, which is usually metallic and is called the *coordination centre*, and a surrounding array of bound molecules or ions, that are in turn known as *ligands* or complexing agents. Many metal-containing compounds, especially those of transition metals, are coordination complexes. A coordination complex whose centre is a metal atom is called a metal complex.

Nomenclature and Terminology

Coordination complexes are so pervasive that their structures and reactions are described in many ways, sometimes confusingly. The atom within a ligand that is bonded to the central metal atom or ion is called the donor atom. In a typical complex, a metal ion is bonded to several donor atoms, which can be the same or different. A polydentate (multiple bonded) ligand is a molecule or ion that bonds to the central atom through several of the ligand's atoms; ligands with 2, 3, 4 or even 6 bonds to the central atom are common. These complexes are called chelate complexes, the formation of such complexes is called chelation, complexation, and coordination.

The central atom or ion, together with all ligands comprise the coordination sphere. The central atoms or ion and the donor atoms comprise the first coordination sphere.

Coordination refers to the "coordinate covalent bonds" (dipolar bonds) between the ligands and the central atom. Originally, a complex implied a reversible association of molecules, atoms, or ions through such weak chemical bonds. As applied to coordination chemistry, this meaning has evolved. Some metal complexes are formed virtually irreversibly and many are bound together by bonds that are quite strong.

The number of donor atoms attached to the central atom or ion is called the coordination number. The most common coordination numbers are 2, 4 and especially 6. A hydrated ion is one kind of a complex ion (or simply a complex), a species formed between a central metal ion and one or more surrounding ligands, molecules or ions that contain at least one lone pair of electrons,

If all the ligands are monodentate, then the number of donor atoms equals the number of ligands. For example, the cobalt(II) hexahydrate ion or the hexaaquacobalt(II) ion $[Co(H_2O)_6]^{2+}$, is a hydrated-complex ion that consists of six water molecules attached to a metal ion Co. The oxidation state and the coordination number reflect the number of bonds formed between the metal ion and the ligands in the complex ion. However the coordination number of $Pt(en)_2^{2+}$ is 4 (rather than 2) since it has two bidentate ligands, which contain four donor atoms in total.

History

Coordination complexes have been known since the beginning of modern chemistry. Early well-known coordination complexes include dyes such as Prussian blue. Their properties were first well understood in the late 1800s, following the 1869 work of Christian Wilhelm Blomstrand. Blomstrand developed what has come to be known as the complex ion chain theory. The theory claimed that the reason coordination complexes form is because in solution, ions would be bound via ammonia chains. He compared this effect to the way that various carbohydrate chains form.

Alfred Werner

Following this theory, Danish scientist Sophus Mads Jorgensen made improvements to it. In his version of the theory, Jorgensen claimed that when a molecule dissociates in a solution there were two possible outcomes: the ions would bind via the ammonia chains Blomstrand had described or the ions would bind directly to the metal.

It was not until 1893 that the most widely accepted version of the theory today was published by Alfred Werner. Werner's work included two important changes to the Blomstrand theory. The first was that Werner described the two different ion possibilities in terms of location in the coordination sphere. He claimed that if the ions were to form a chain this would occur outside of the coordination sphere while the ions that bound directly to the metal would do so within the coordination sphere. In one of Werner's most important discoveries however he disproved the majority of the chain theory. Werner was able to discover the spatial arrangements of the ligands that were involved in the formation of the complex hexacoordinate cobalt. His theory allows one to understand the difference between a coordinated ligand and a charge balancing ion in a compound, for example the chloride ion in the cobaltammine chlorides and to explain many of the previously inexplicable isomers.

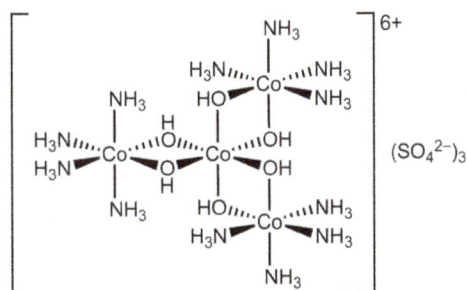

Structure of hexol

In 1914, Werner first resolved the coordination complex, called hexol, into optical isomers, overthrowing the theory that only carbon compounds could possess chirality.

Structures

The ions or molecules surrounding the central atom are called ligands. Ligands are generally bound to the central atom by a coordinate covalent bond (donating electrons from a lone electron pair into an empty metal orbital), and are said to be coordinated to the atom. There are also organic ligands such as alkenes whose pi bonds can coordinate to empty metal orbitals. An example is ethene in the complex known as Zeise's salt, $K^+[PtCl_3(C_2H_4)]^-$.

Geometry

In coordination chemistry, a structure is first described by its coordination number, the number of ligands attached to the metal (more specifically, the number of donor atoms). Usually one can count the ligands attached, but sometimes even the counting can become ambiguous. Coordination numbers are normally between two and nine, but large numbers of ligands are not uncommon for the lanthanides and actinides. The number of bonds depends on the size, charge, and electron configuration of the metal ion and the ligands. Metal ions may have more than one coordination number.

Typically the chemistry of transition metal complexes is dominated by interactions between s and p molecular orbitals of the ligands and the d orbitals of the metal ions. The s, p, and d orbitals of the metal can accommodate 18 electrons. The maximum coordination number for a certain metal is thus related to the electronic configuration of the metal ion (to be more specific, the number of empty orbitals) and to the ratio of the size of the ligands and the metal ion. Large metals and small ligands lead to high coordination numbers, e.g. $[Mo(CN)_8]^{4-}$. Small metals with large ligands lead to low coordination numbers, e.g. $Pt[P(CMe_3)]_2$. Due to their large size, lanthanides, actinides, and early transition metals tend to have high coordination numbers.

Different ligand structural arrangements result from the coordination number. Most structures follow the points-on-a-sphere pattern (or, as if the central atom were in the middle of a polyhedron where the corners of that shape are the locations of the ligands), where orbital overlap (between ligand and metal orbitals) and ligand-ligand repulsions tend to lead to certain regular geometries. The most observed geometries are listed below. There are cases that deviate from a regular geometry due to the use of ligands of different types (which results in irregular bond lengths) or due to the size of ligands.

- Linear for two-coordination

- Trigonal planar for three-coordination

- Tetrahedral or square planar for four-coordination

- Trigonal bipyramidal or square pyramidal for five-coordination

- Octahedral (orthogonal) for six-coordination

- Pentagonal bipyramidal, capped octahedral or capped trigonal prismatic for seven-coordination

- Square antiprismatic or dodecahedral for eight-coordination

- Tri-capped trigonal prismatic (triaugmented triangular prism) or capped square antiprismatic for nine-coordination.

Due to special electronic effects such as (second-order) Jahn–Teller stabilization, certain geometries (in which the coordination atoms do not follow a points-on-a-sphere pattern) are stabilized relative to the other possibilities, e.g. for some compounds the trigonal prismatic geometry is stabilized relative to octahedral structures for six-coordination.

- Bent for two-coordination

- Trigonal pyramidal for three-coordination

- Trigonal prismatic for six-coordination

Isomerism

The arrangement of the ligands is fixed for a given complex, but in some cases it is mutable by a reaction that forms another stable isomer.

There exist many kinds of isomerism in coordination complexes, just as in many other compounds.

Stereoisomerism

Stereoisomerism occurs with the same bonds in different orientations relative to one another. Stereoisomerism can be further classified into:

Cis–trans isomerism and Facial–meridional Isomerism

Cis–trans isomerism occurs in octahedral and square planar complexes (but not tetrahedral). When two ligands are adjacent they are said to be cis, when opposite each other, trans. When three identical ligands occupy one face of an octahedron, the isomer is said to be facial, or fac. In a *fac* isomer, any two identical ligands are adjacent or *cis* to each other. If these three ligands and the metal ion are in one plane, the isomer is said to be meridional, or mer. A *mer* isomer can be considered as a combination of a *trans* and a *cis*, since it contains both trans and cis pairs of identical ligands.

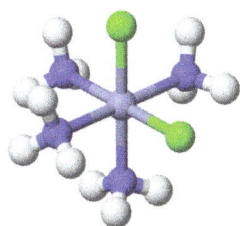

cis-[CoCl$_2$(NH$_3$)$_4$]$^+$ *trans*-[CoCl$_2$(NH$_3$)$_4$]$^+$ *fac*-[CoCl$_3$(NH$_3$)$_3$]   *mer*-[CoCl$_3$(NH$_3$)$_3$]

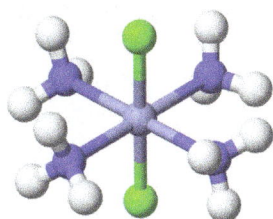

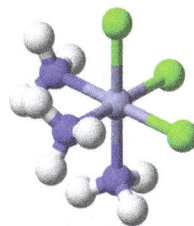

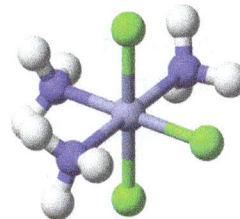

Optical Isomerism

Optical isomerism occurs when a molecule is not superimposable with its mirror image. It is so called because the two isomers are each optically active, that is, they rotate the plane of polarized light in opposite directions. The symbol Λ (*lambda*) is used as a prefix to describe the left-handed propeller twist formed by three bidentate ligands, as shown. Likewise, the symbol Δ (*delta*) is used as a prefix for the right-handed propeller twist.

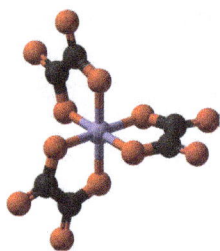

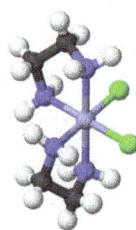

Λ-[Fe(ox)$_3$]$^{3-}$ Δ-[Fe(ox)$_3$]$^{3-}$ Λ-*cis*-[CoCl$_2$(en)$_2$]$^+$ Δ-*cis*-[CoCl$_2$(en)$_2$]$^+$

Structural Isomerism

Structural isomerism occurs when the bonds are themselves different. There are four types of structural isomerism: ionisation isomerism, solvate or hydrate isomerism, linkage isomerism and coordination isomerism.

1. Ionisation isomerism – the isomers give different ions in solution although they have the same composition. This type of isomerism occurs when the counter ion of the complex is also a potential ligand. For example, pentaamminebromocobalt(III) sulfate [Co(NH$_3$)$_5$Br]SO$_4$ is red violet and in solution gives a precipitate with barium chloride, confirming the presence of sulfate ion, while pentaamminesulfatecobalt(III) bromide [Co(NH$_3$)$_5$SO$_4$]Br is red and tests negative for sulfate ion in solution, but instead gives a precipitate of AgBr with silver nitrate.

2. Solvate or hydrate isomerism – the isomers have the same composition but differ with respect to the number of solvent ligand molecules as well as the counter ion in the crystal lattice. For example, [Cr(H$_2$O)$_6$]Cl$_3$ is violet colored, [CrCl(H$_2$O)$_5$]Cl$_2$·H$_2$O is blue-green, and [CrCl$_2$(H$_2$O)$_4$]Cl·2H$_2$O is dark green.

3. Linkage isomerism occurs with ambidentate ligands that can bind in more than one place. For example, NO$_2$ is an ambidentate ligand: It can bind to a metal at either the N atom or an O atom.

4. Coordination isomerism – this occurs when both positive and negative ions of a salt are complex ions and the two isomers differ in the distribution of ligands between the cation and the anion. For example, [Co(NH$_3$)$_6$][Cr(CN)$_6$] and [Cr(NH$_3$)$_6$][Co(CN)$_6$].

Electronic Properties

Many of the properties of transition metal complexes are dictated by their electronic structures. The electronic structure can be described by a relatively ionic model that ascribes formal charges to the metals and ligands. This approach is the essence of crystal field theory (CFT). Crystal field theory, introduced by Hans Bethe in 1929, gives a quantum mechanically based attempt at understanding complexes. But crystal field theory treats all interactions in a complex as ionic and assumes that the ligands can be approximated by negative point charges.

More sophisticated models embrace covalency, and this approach is described by ligand field theory (LFT) and Molecular orbital theory (MO). Ligand field theory, introduced in 1935 and built from molecular orbital theory, can handle a broader range of complexes and can explain complexes in which the interactions are covalent. The chemical applications of group theory can aid in the understanding of crystal or ligand field theory, by allowing simple, symmetry based solutions to the formal equations.

Chemists tend to employ the simplest model required to predict the properties of interest; for this reason, CFT has been a favorite for the discussions when possible. MO and LF theories are more complicated, but provide a more realistic perspective.

The electronic configuration of the complexes gives them some important properties:

Color of Transition Metal Complexes

Synthesis of copper(II)-tetraphenylporphyrin, a metal complex, from tetraphenylporphyrin and copper(II) acetate monohydrate.

Transition metal complexes often have spectacular colors caused by electronic transitions by the absorption of light. For this reason they are often applied as pigments. Most transitions that are related to colored metal complexes are either d–d transitions or charge transfer bands. In a d–d transition, an electron in a d orbital on the metal is excited by a photon to another d orbital of higher energy. A charge transfer band entails promotion of an electron from a metal-based orbital into an empty ligand-based orbital (Metal-to-Ligand Charge Transfer or MLCT). The converse also occurs: excitation of an electron in a ligand-based orbital into an empty metal-based orbital (Ligand to Metal Charge Transfer or LMCT). These phenomena can be observed with the aid of electronic spectroscopy; also known as UV-Vis. For simple compounds with high symmetry, the d–d transitions can be assigned using Tanabe–Sugano diagrams. These assignments are gaining increased support with computational chemistry.

Colours of Various Example Coordination Complexes						
	Fe^{2+}	Fe^{3+}	Co^{2+}	Cu^{2+}	Al^{3+}	Cr^{3+}
Hydrated Ion	$[Fe(H_2O)_6]^{2+}$ Pale green Solution	$[Fe(H_2O)_6]^{3+}$ Yellow/brown Solution	$[Co(H_2O)_6]^{2+}$ Pink Solution	$[Cu(H_2O)_6]^{2+}$ Blue Solution	$[Al(H_2O)_6]^{3+}$ Colourless Solution	$[Cr(H_2O)_6]^{3+}$ Green Solution
OH^-, dilute	$[Fe(H_2O)_4(OH)_2]$ Dark green Precipitate	$[Fe(H_2O)_3(OH)_3]$ Brown Precipitate	$[Co(H_2O)_4(OH)_2]$ Blue/green Precipitate	$[Cu(H_2O)_4(OH)_2]$ Blue Precipitate	$[Al(H_2O)_3(OH)_3]$ White Precipitate	$[Cr(H_2O)_3(OH)_3]$ Green Precipitate
OH^-, concentrated	$[Fe(H_2O)_4(OH)_2]$ Dark green Precipitate	$[Fe(H_2O)_3(OH)_3]$ Brown Precipitate	$[Co(H_2O)_4(OH)_2]$ Blue/green Precipitate	$[Cu(H_2O)_4(OH)_2]$ Blue Precipitate	$[Al(OH)_4]^-$ Colourless Solution	$[Cr(OH)_6]^{3-}$ Green Solution

NH₃, dilute	$[Fe(H_2O)_4(OH)_2]$ Dark green Precipitate	$[Fe(H_2O)_3(OH)_3]$ Brown Precipitate	$[Co(H_2O)_4(OH)_2]$ Blue/green Precipitate	$[Cu(H_2O)_4(OH)_2]$ Blue Precipitate	$[Al(H_2O)_3(OH)_3]$ White Precipitate	$[Cr(H_2O)_3(OH)_3]$ Green Precipitate
NH₃, concentrated	$[Fe(H_2O)_4(OH)_2]$ Dark green Precipitate	$[Fe(H_2O)_3(OH)_3]$ Brown Precipitate	$[Co(NH_3)_6]^{2+}$ Straw coloured Solution	$[Cu(NH_3)_4(H_2O)_2]^{2+}$ Deep blue Solution	$[Al(H_2O)_3(OH)_3]$ White Precipitate	$[Cr(NH_3)_6]^{3+}$ Purple Solution
CO_3^{2-}	$FeCO_3$ Dark green Precipitate	$[Fe(H_2O)_3(OH)_3]$ Brown Precipitate + bubbles	$CoCO_3$ Pink Precipitate	$CuCO_3$ Blue/green Precipitate		

Colors of Lanthanide Complexes

Superficially lanthanide complexes are similar to those of the transition metals in that some are coloured. However, for the common Ln^{3+} ions (Ln = lanthanide) the colors are all pale, and hardly influenced by the nature of the ligand. The colors are due to 4f electron transitions. As the 4f orbitals in lanthanides are "buried" in the xenon core and shielded from the ligand by the 5s and 5p orbitals they are therefore not influenced by the ligands to any great extent leading to a much smaller crystal field splitting than in the transition metals. The absorption spectra of an Ln^{3+} ion approximates to that of the free ion where the electronic states are described by spin-orbit coupling (also called L-S coupling or Russell-Saunders coupling). This contrasts to the transition metals where the ground state is split by the crystal field. Absorptions for Ln^{3+} are weak as electric dipole transitions are parity forbidden (Laporte Rule forbidden) but can gain intensity due to the effect of a low-symmetry ligand field or mixing with higher electronic states (*e.g.* d orbitals). Also absorption bands are extremely sharp which contrasts with those observed for transition metals which generally have broad bands. This can lead to extremely unusual effects, such as significant color changes under different forms of lighting.

Magnetism

Metal complexes that have unpaired electrons are magnetic. Considering only monometallic complexes, unpaired electrons arise because the complex has an odd number of electrons or because electron pairing is destabilized. Thus, monomeric Ti(III) species have one "d-electron" and must be (para)magnetic, regardless of the geometry or the nature of the ligands. Ti(II), with two d-electrons, forms some complexes that have two unpaired electrons and others with none. This effect is illustrated by the compounds $TiX_2[(CH_3)_2PCH_2CH_2P(CH_3)_2]_2$: when X = Cl, the complex is paramagnetic (high-spin configuration), whereas when X = CH_3, it is diamagnetic (low-spin configuration). It is important to realize that ligands provide an important means of adjusting the ground state properties.

In bi- and polymetallic complexes, in which the individual centres have an odd number of electrons or that are high-spin, the situation is more complicated. If there is interaction (either direct or through ligand) between the two (or more) metal centres, the electrons may couple (antiferromagnetic coupling, resulting in a diamagnetic compound), or they may enhance each other (ferromagnetic coupling). When there is no interaction, the two (or more) individual metal centers behave as if in two separate molecules.

Reactivity

Complexes show a variety of possible reactivities:

- Electron transfers

  A common reaction between coordination complexes involving ligands are inner and outer sphere electron transfers. They are two different mechanisms of electron transfer redox reactions, largely defined by the late Henry Taube. In an inner sphere reaction, a ligand with two lone electron pairs acts as a *bridging ligand*, a ligand to which both coordination centres can bond. Through this, electrons are transferred from one centre to another.

- (Degenerate) ligand exchange

  One important indicator of reactivity is the rate of degenerate exchange of ligands. For example, the rate of interchange of coordinate water in $[M(H_2O)_6]^{n+}$ complexes varies over 20 orders of magnitude. Complexes where the ligands are released and rebound rapidly are classified as labile. Such labile complexes can be quite stable thermodynamically. Typical labile metal complexes either have low-charge (Na^+), electrons in d-orbitals that are antibonding with respect to the ligands (Zn^{2+}), or lack covalency (Ln^{3+}, where Ln is any lanthanide). The lability of a metal complex also depends on the high-spin vs. low-spin configurations when such is possible. Thus, high-spin Fe(II) and Co(III) form labile complexes, whereas low-spin analogues are inert. Cr(III) can exist only in the low-spin state (quartet), which is inert because of its high formal oxidation state, absence of electrons in orbitals that are M−L antibonding, plus some "ligand field stabilization" associated with the d^3 configuration.

- Associative processes

  Complexes that have unfilled or half-filled orbitals often show the capability to react with substrates. Most substrates have a singlet ground-state; that is, they have lone electron pairs (e.g., water, amines, ethers), so these substrates need an empty orbital to be able to react with a metal centre. Some substrates (e.g., molecular oxygen) have a triplet ground state, which results that metals with half-filled orbitals have a tendency to react with such substrates (it must be said that the dioxygen molecule also has lone pairs, so it is also capable to react as a 'normal' Lewis base).

If the ligands around the metal are carefully chosen, the metal can aid in (stoichiometric or catalytic) transformations of molecules or be used as a sensor.

Classification

Metal complexes, also known as coordination compounds, include all metal compounds, aside from metal vapors, plasmas, and alloys. The study of "coordination chemistry" is the study of "inorganic chemistry" of all alkali and alkaline earth metals, transition metals, lanthanides, actinides, and metalloids. Thus, coordination chemistry is the chemistry of the majority of the periodic table. Metals and metal ions exist, in the condensed phases at least, only surrounded by ligands.

The areas of coordination chemistry can be classified according to the nature of the ligands, in broad terms:

- Classical (or "Werner Complexes"): Ligands in classical coordination chemistry bind to metals, almost exclusively, via their "lone pairs" of electrons residing on the main group atoms of the ligand. Typical ligands are H_2O, NH_3, Cl^-, CN^-, en. Some of the simplest members of such complexes are described in metal aquo complexes, metal ammine complexes,

  Examples: $[Co(EDTA)]^-$, $[Co(NH_3)_6]Cl_3$, $[Fe(C_2O_4)_3]K_3$

- Organometallic Chemistry: Ligands are organic (alkenes, alkynes, alkyls) as well as "organic-like" ligands such as phosphines, hydride, and CO.

  Example: $(C_5H_5)Fe(CO)_2CH_3$

- Bioinorganic Chemistry: Ligands are those provided by nature, especially including the side chains of amino acids, and many cofactors such as porphyrins.

  Example: hemoglobin contains heme, a porphyrin complex of iron

  Example: chlorophyll contains a porphyrin complex of magnesium

  Many natural ligands are "classical" especially including water.

- Cluster Chemistry: Ligands are all of the above also include other metals as ligands.

  Example $Ru_3(CO)_{12}$

- In some cases there are combinations of different fields:

  Example: $[Fe_4S_4(Scysteinyl)_4]^{2-}$, in which a cluster is embedded in a biologically active species.

Mineralogy, materials science, and solid state chemistry – as they apply to metal ions – are subsets of coordination chemistry in the sense that the metals are surrounded by ligands. In many cases these ligands are oxides or sulfides, but the metals are coordinated nonetheless, and the principles and guidelines discussed below apply. In hydrates, at least some of the ligands are water molecules. It is true that the focus of mineralogy, materials science, and solid state chemistry differs from the usual focus of coordination or inorganic chemistry. The former are concerned primarily with polymeric structures, properties arising from a collective effects of many highly interconnected metals. In contrast, coordination chemistry focuses on reactivity and properties of complexes containing individual metal atoms or small ensembles of metal atoms.

Naming Complexes

The basic procedure for naming a complex:

1. When naming a complex ion, the ligands are named before the metal ion.

2. Write the names of the ligands in alphabetical order. (Numerical prefixes do not affect the order.)

   o Multiple occurring monodentate ligands receive a prefix according to the number of occurrences: *di-*, *tri-*, *tetra-*, *penta-*, or *hexa*. Polydentate ligands (e.g., ethylenediamine, oxalate) receive *bis-*, *tris-*, *tetrakis-*, etc.

   o Anions end in *o*. This replaces the final 'e' when the anion ends with '-ide', '-ate' or '-ite', e.g. *chloride* becomes *chlorido* and *sulfate* becomes *sulfato*. Formerly, '-ide' was changed to '-o' (e.g. *chloro* and *cyano*), but this rule has been modified in the 2005 IUPAC recommendations and the correct forms for these ligands are now *chlorido* and *cyanido*.

o Neutral ligands are given their usual name, with some exceptions: NH_3 becomes *ammine*; H_2O becomes *aqua* or *aquo*; CO becomes *carbonyl*; NO becomes *nitrosyl*.

3. Write the name of the central atom/ion. If the complex is an anion, the central atom's name will end in *-ate*, and its Latin name will be used if available (except for mercury).

4. If the central atom's oxidation state needs to be specified (when it is one of several possible, or zero), write it as a Roman numeral (or 0) in parentheses.

5. Name cation then anion as separate words (if applicable, as in last example)

Examples:

metal	changed to
cobalt	cobaltate
aluminium	aluminate
chromium	chromate
vanadium	vanadate
copper	cuprate
iron	ferrate

$[NiCl_4]^{2-}$ → tetrachloronickelate(II) ion

$[CuCl_5NH_3]^{3-}$ → amminepentachlorocuprate(II) ion

$[Cd(CN)_2(en)_2]$ → dicyanobis(ethylenediamine)cadmium(II)

$[CoCl(NH_3)_5]SO_4$ → pentaamminechlorocobalt(III) sulfate

The coordination number of ligands attached to more than one metal (bridging ligands) is indicated by a subscript to the Greek symbol μ placed before the ligand name. Thus the dimer of aluminium trichloride is described by $Al_2Cl_4(\mu_2\text{-}Cl)_2$.

Stability Constant

The affinity of metal ions for ligands is described by stability constant. This constant, also referred to as the formation constant, is given the notation of K_f and can be calculated through the following method for simple cases:

$$(X)\,Metal_{(aq)} + (Y)\,Lewis\;Base_{(aq)} = (Z)\,Complex\;Ion_{(aq)}$$

$$K_f = \frac{[Complex\;ion]^Z}{[Metal\;ion]^X [Lewis\;base]^Y}$$

Formation constants vary widely. Large values indicate that the metal has high affinity for the ligand, provided the system is at equilibrium.

Sometimes the stability constant will be in a different form known as the constant of destability. This constant is expressed as the inverse of the constant of formation and is denoted as $K_d = 1/K_f$. This constant represents the reverse reaction for the decomposition of a complex ion into its individual metal and ligand components. When comparing the values for K_d, the larger the value is the more unstable the complex ion is.

As a result of these complex ions forming in solutions they also can play a key role in solubility of other compounds. When a complex ion is formed it can alter the concentrations of its components in the solution. For example:

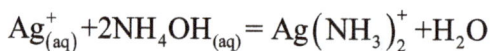

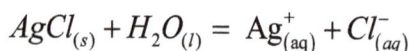

$$Ag^+_{(aq)} + 2NH_4OH_{(aq)} = Ag(NH_3)_2^+ + H_2O$$

$$AgCl_{(s)} + H_2O_{(l)} = Ag^+_{(aq)} + Cl^-_{(aq)}$$

In these reactions which both occurred in the same reaction vessel, the solubility of the silver chloride would be increased as a result of the formation of the complex ion. The complex ion formation is favorable takes away a significant portion of the silver ions in solution, as a result the equilibrium for the formation of silver ions from silver chloride will shift to the right to make up for the deficit.

This new solubility can be calculated given the values of K_f and K_{sp} for the original reactions. The solubility is found essentially by combining the two separate equilibria into one combined equilibrium reaction and this combined reaction is the one that determines the new solubility. So K_c, the new solubility constant, is denoted by $K_c = K_{sp} * K_f$.

Application of Coordination Compounds

Metals only exist in solution as coordination complexes, it follows then that this class of compounds is useful in a wide variety of ways. Coordination compounds are therefore found both in nature and in industry (in, especially, color-rich ways). Some common complex ions include such substances as vitamin B_{12}, the heme group in hemoglobin and the cytochromes, and the chlorin group in chlorophyll (which are dark red or cherry colored, blood red, and green in color respectively), and some dyes and pigments. One major use of coordination compounds is in homogeneous catalysis for the production of organic substances.

Coordination compounds have uses in both nature and in industry. Coordination compounds are vital to many living organisms. For example, many enzymes are metal complexes, like carboxypeptidase, a hydrolytic enzyme important in digestion. This enzyme consists of a zinc ion surrounded by many amino acid residues. Another complex ion enzyme is catalase, which decomposes the cell's waste hydrogen peroxide. This enzyme contains iron-porphyrin complexes, similar to those of heme in the hemoglobin molecule. Chlorophyll contains a magnesium-porphyrin complexes (chlorin), and vitamin B_{12} is a complex with cobalt and corrin.

Coordination compounds are also widely used in industry. The intense colors of many compounds render them of great use as dyes and pigments. Specifically phthalocyanine complexes are an important class of dyes for fabrics. Nickel, cobalt, and copper can be extracted using hydrometallurgical processes involving complex ions. They are extracted from their ores as ammine complexes with aqueous ammonia. Metals can also be separated using the selective precipitation and solubility of complex ions, as explained in later paragraphs. Cyanide complexes are often used in electroplating.

Coordination compounds can also be used to identify unknown substances in a solution. This analysis can be done by utilizing the selective precipitation of the complex ions, the formation of color complexes which can be measured spectrophotometrically, or the preparation of complexes, such as metal acetylacetonates, which can be separated with organic solvents.

A combination of titanium trichloride and triethylaluminum brings about the polymerization of organic compounds with carbon-carbon double bonds to form polymers of high molecular weight and ordered structures. Many of these polymers are of great commercial importance because they are used in common fibers, films, and plastics.

Other common uses of coordination compounds in industry include the following:

1. They are used in photography, i.e., AgBr forms a soluble complex with sodium thiosulfate in photography.

2. $K[Ag(CN)_2]$ is used forelectroplating of silver, and $K[Au(CN)_2]$ is used for gold plating.

3. Some ligands oxidise Co^{2+} to Co^{3+} ion.

4. Ethylenediaminetetraacetic acid (EDTA) is used for estimation of Ca^{2+} and Mg^{2+} in hard water.

5. Silver and gold are extracted by treating zinc with their cyanide complexes

Molecular Recognition in Supramolecular Systems

Chemists have demonstrated that artificial supramolecular systems can be designed that exhibit molecular recognition. One of the earliest examples of such a system is crown ethers which are capable of selectively binding specific cations. However, a number of artificial systems have since been established. "For their development and use of molecules with structure-specific interactions of high selectivity," Charles Pedersen, Jean-Marie Lehn, and Donald Cram have received the 1987 Nobel prize in chemistry.

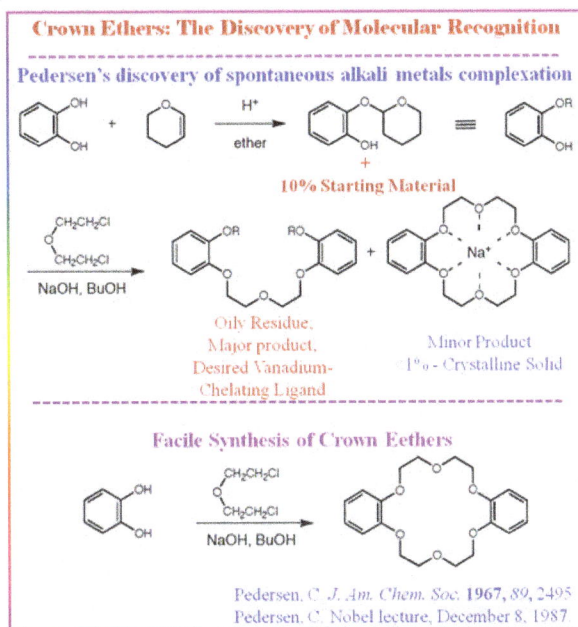

Pedersen's synthesis of crown ether and the discovery of Molecular recognition.

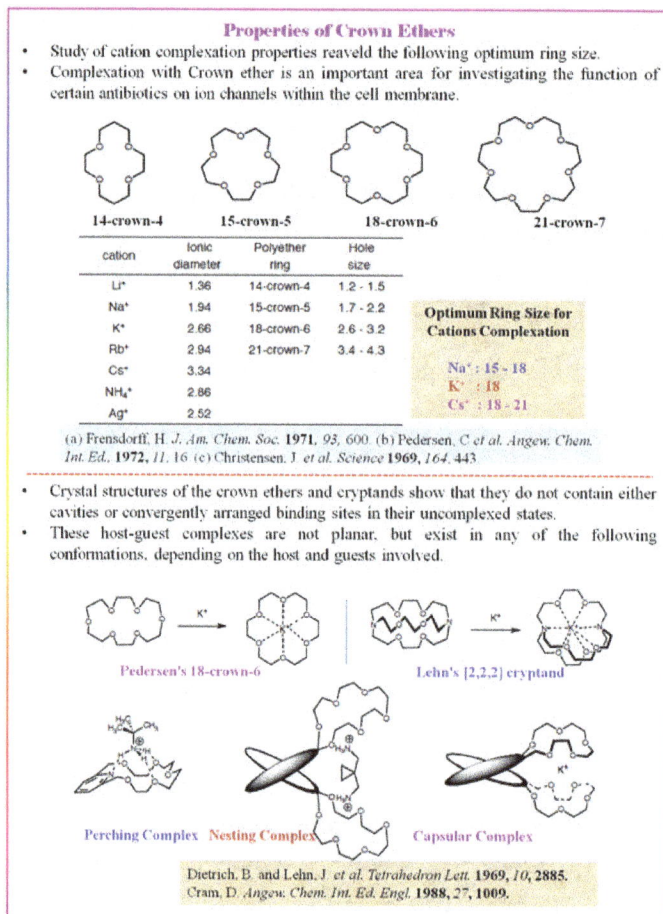

Properties of crown ethers.

Applications of Molecular Recognition

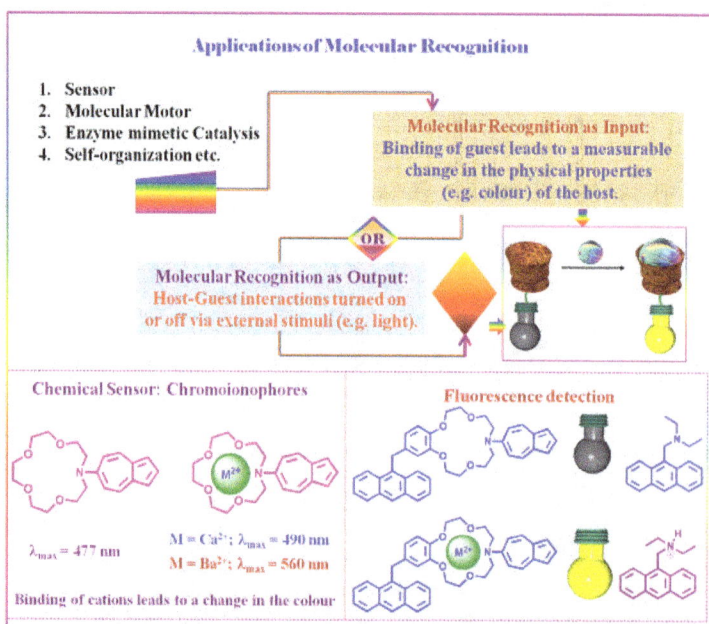

Enantiomer Differentiation and the Crown Ether

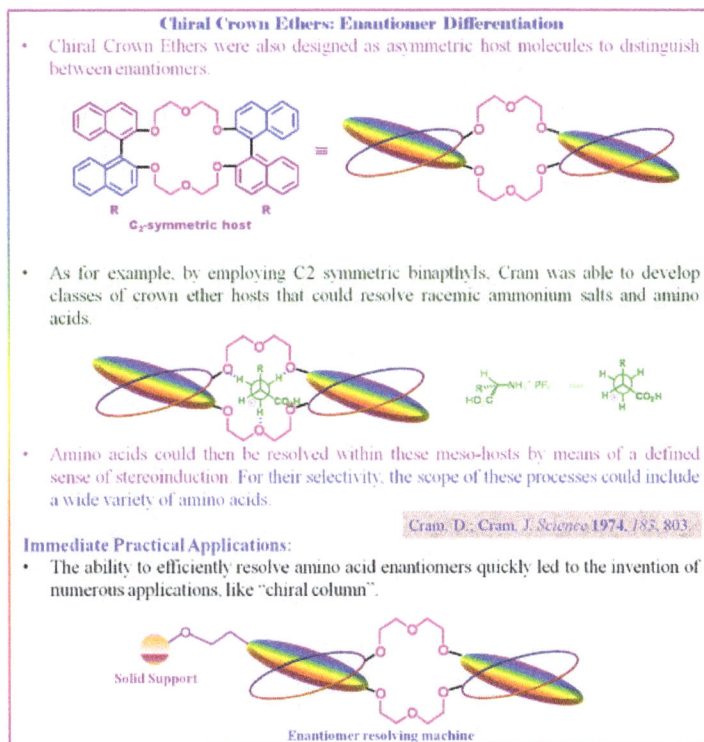

Molecular Recognition of Ammonium Ions

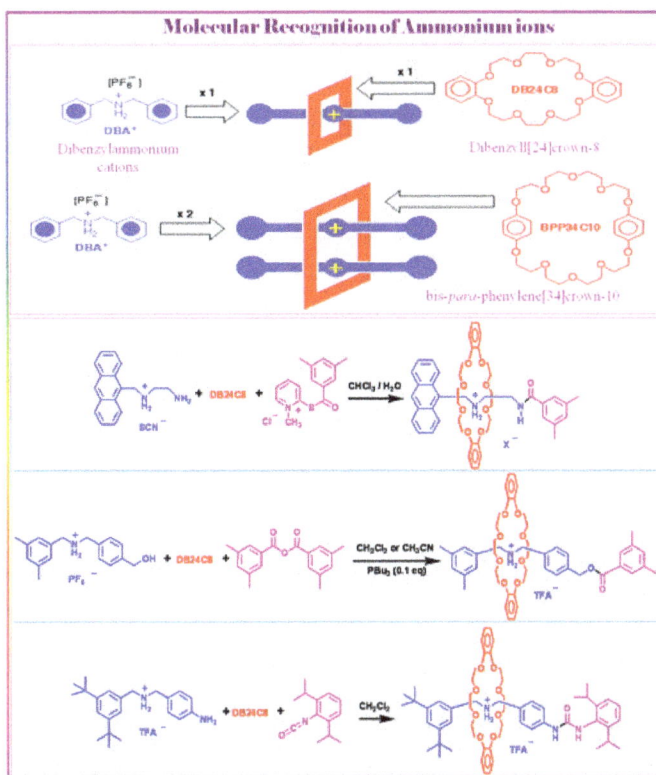

Encapsulation Era

The research in molecular recognition has progressed far beyond the sequestration of ions by macrocyclic polyether like crown ethers during the past two decades. Therefore, many Hosts have been designed and synthesized with a variety of shapes for binding charged or neutral Guests. All of these Hosts share a common feature of having concave surfaces to accommodate convex guests. The next generation of hosts was designed to encounter all possible guests' surfaces with high selectivity. Recently, superstructure generated by multiple copies of small molecules through self-recognition via weak intermolecular forces surrounding a target was also developed which is called a reversible encapsulation.

Characteristics of Encapsulation Complexes

a. They are synthetic, self-assembled hosts that surround their guest molecules.

b. They are dynamic and form reversibly in solution with varying range of lifetimes.

c. The capsules isolate molecules from the bulk solution.

d. They self-assemble only in the presence of suitable guests to give encapsulation complexes.

Uses of Encapsulation Complexes

a. as a reaction chamber for different guests inside the capsules

b. Capsule for a chiral receptor.

c. Cavity for single-molecule solvation can be observed

d. used to stabilize reactive intermediates and transition states

e. Used to alter the course of reactions.

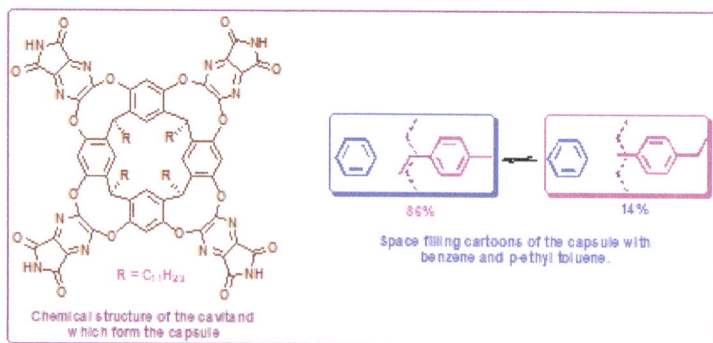

Cavitant and the capsule formation.

For the capsule formed by the molecule X in figure Y exists as two complexes in presence of benzene and p-ethyl toluene. The two molecules are too large to slip past each other, and the p-ethyl toluene is too long to tumble freely while inside the capsule. Two benzene molecules as guests can gather the capsule and filled 41% of the space. With p-ethyl toluene, one molecule fills only 33% of the space, but two molecules are too long to be accommodated in the cavity. However maximum space filling (53% of the space) is observed in a combination of benzene/p-ethyl toluene which the author says as "optimal filling of the capsule's space". This matching of host space and guest size drives much encapsulation and is especially good for arranging bimolecular reactions inside.

Frontires of Encapsulation Era

Frontiers of Encapsulation Era:

- Advances in predictive modeling technology as well as 2D NMR techniques have propelled progress in this area far beyond the days of mere molecular recognition
- By employing arrays of purposefully positioned H-bond donors and acceptors, researchers have been able to "synthesize" the large, self-assembling structures (>400 Å)
- These self-assembling capsules employ the same non-covalent interactions to bring together multiple guest molecules within these hosts.
- These self-assembling capsules employ the same non-covalent interactions to bring together multiple guest molecules within these hosts.

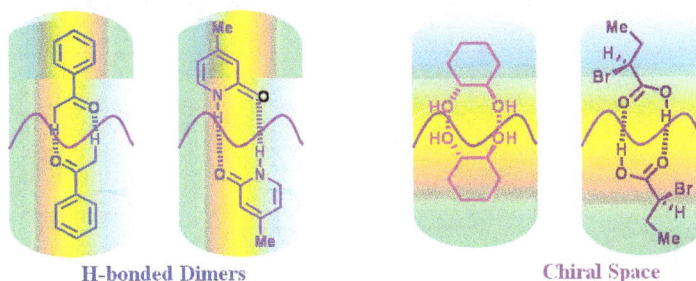

H-bonded Dimers   Chiral Space

Rebek, J. Angew. Chem. Int. Ed. Engl. 2005, 44, 2068.

- When complementary guests assemble together within a capsule, accellerated reactivity can occur in a similar manner to the induced proximity effect created within an enzyme.

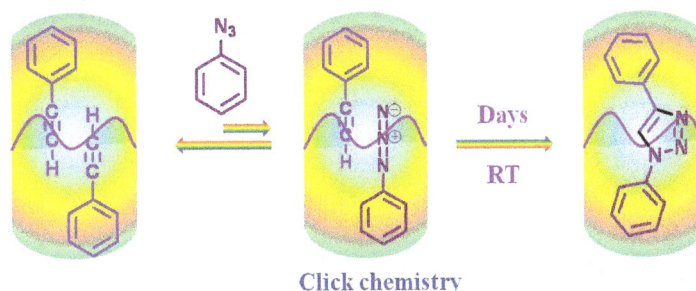

Click chemistry

Molecular Recognition in Biological Systems

Molecular recognition plays an important role in biological systems and is observed in between receptor-ligand, antigen-antibody, DNA-protein, sugar-lectin, RNA-ribosome, etc. An important example of molecular recognition is the antibiotic vancomycin that selectively binds with the peptides with terminal D-alanyl-D-alanine in bacterial cells through five hydrogen bonds. The vancomycin is lethal to the bacteria since once it has bound to these particular peptides they are unable to be used to construct the bacteria's cell wall.

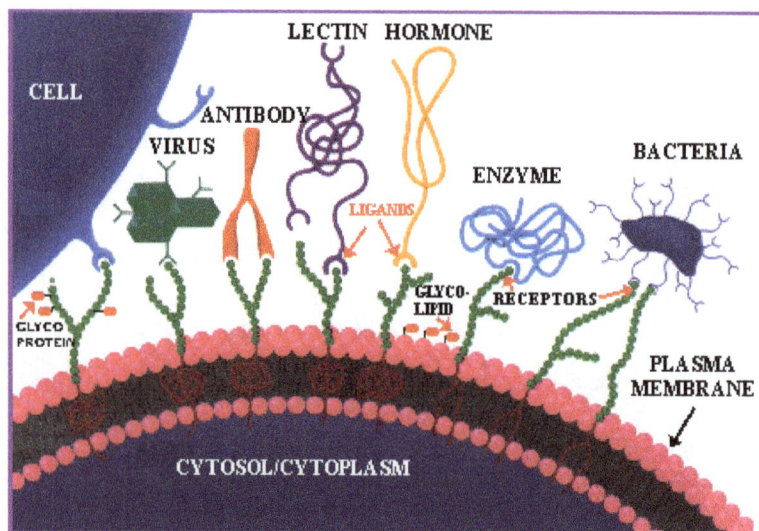

Various molecules exist upon molecular recognition in a cell.

Molecular Recognition in Biological Systems

- Non covalent interactions between different sets of macromolecules leads to supramolecular assembiles that serve specific subcellular functions in biology.
- For example, ribosomes are supramolecular assemblies of proteins and RNA.

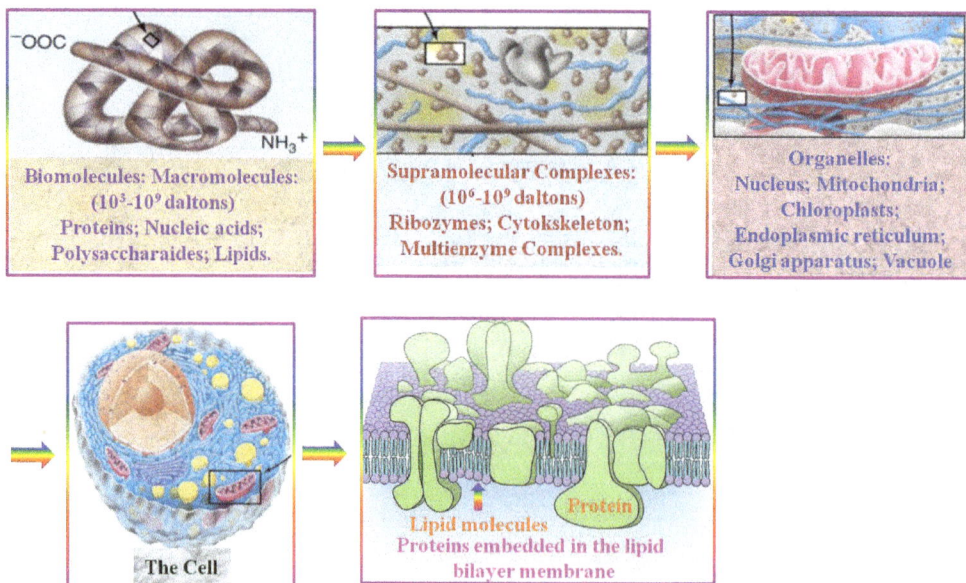

Biomolecules: Macromolecules:
(10⁵-10⁹ daltons)
Proteins; Nucleic acids;
Polysaccharaides; Lipids.

Supramolecular Complexes:
(10⁶-10⁹ daltons)
Ribozymes; Cytokskeleton;
Multienzyme Complexes.

Organelles:
Nucleus; Mitochondria;
Chloroplasts;
Endoplasmic reticulum;
Golgi apparatus; Vacuole

The Cell

Lipid molecules
Proteins embedded in the lipid bilayer membrane
Protein

- Packing of supramolecular assemblies into cellular inclusions surrounded by membranes, called organelles. Organelles are dedicated to specific cellular tasks. Organelles are present only in higher organisms (eukaryotes).
- Membranes are complexes of proteins and lipid molecules maintained by *non covalent forces*.
- Hydrophobic interactions are particularly important in maintaining membrane structure.
- The spontaneous assembly of membranes in an aqueous environment is the result of hydrophobicity of the membrane lipids and proteins.

Interactions and recognition of various molecules in a cell.

Protein-Ligand Complexation-A Molecular Recognition

Molecular Recognition in Biological Systems: Protein-Ligand Complexation

- Changing structure of ligand leads to changes in binding affinity.
- ligand preorganization, hydrophobicity, hydrogen bonding capability, π-cation stabilizing ability, etc. affect energetics in protein-ligand interactions

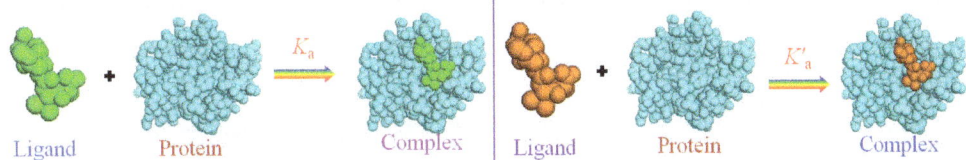

| Ligand | Protein | K_a | Complex | Ligand | Protein | K'_a | Complex |

- Optimizing protein binding affinity is critical first step in drug discovery.

General Features of Protein-Ligand Interactions

- Changes in rotational and translational degrees of freedom.
- Shape complementarity of binding surfaces, but conformational changes occur.
- Hydrating water molecules around protein and ligand reorganize and some are released to bulk water – desolvation.
- Formation of non-bonded interactions i.e. protein-ligand complex.

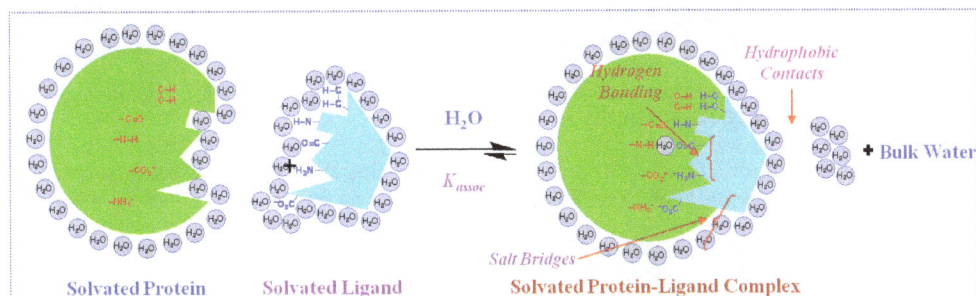

Solvated Protein   Solvated Ligand   Solvated Protein-Ligand Complex

Ligand Affinity of Proteins Increases by Pre-organizing Ligands

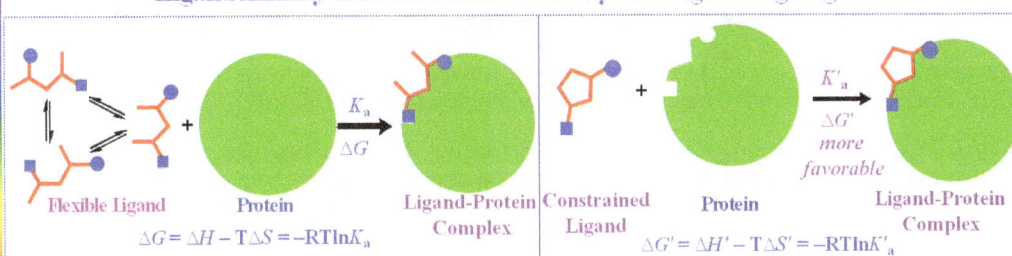

Flexible Ligand   Protein   Ligand-Protein Complex   Constrained Ligand   Protein   Ligand-Protein Complex

$$\Delta G = \Delta H - T\Delta S = -RT\ln K_a$$

$$\Delta G' = \Delta H' - T\Delta S' = -RT\ln K'_a$$

- There are various benefits of preorganization in protein-ligand interactions-like ligand binding is strong.
- $\Delta G'$ should be more negative than ΔG because $\Delta S'$ less negative (more positive) than ΔS in this types of reactions inside the proteins.

References

- Wiley, G.R.; Miller, S.I. (1972). "Thermodynamic parameters for hydrogen bonding of chloroform with Lewis bases in cyclohexane. Proton magnetic resonance study". Journal of the American Chemical Society. 94 (10): 3287. doi:10.1021/ja00765a001

- Anslyn, Eric (2004). Modern Physical Organic Chemistry. Sausalito, CA: University Science. ISBN 978-1-891389-31-3

- Cockroft, Scott L.; Hunter, Christopher A. (1 January 2007). "Chemical double-mutant cycles: dissecting non-covalent interactions". Chemical Society Reviews. 36 (2): 172. doi:10.1039/b603842p

- Kwak, K; Rosenfeld, DE; Chung, JK; Fayer, MD (2008). "Solute-solvent complex switching dynamics of chloroform between acetone and dimethylsulfoxide-two-dimensional IR chemical exchange spectroscopy". The Journal of Physical Chemistry B. 112 (44): 13906–15. PMC 2646412 . PMID 18855462. doi:10.1021/jp806035w

- Brown, Theodore; et al. (2009). Chemistry : the central science (11th ed.). Upper Saddle River, NJ: Pearson Prentice Hall. ISBN 978-0-13-600617-6

- John R. Sabin (1971). "Hydrogen bonds involving sulfur. I. Hydrogen sulfide dimer". J. Am. Chem. Soc. 93 (15): 3613–3620. doi:10.1021/ja00744a012

- Grunenberg, Jörg (2004). "Direct Assessment of Interresidue Forces in Watson–Crick Base Pairs Using Theoretical Compliance Constants". Journal of the American Chemical Society. 126 (50): 16310–1. PMID 15600318. doi:10.1021/ja046282a

- Silverman, Richard B. (2004). The organic chemistry of drug design and drug action (2. ed.). Amsterdam [u.a.]: Elsevier. ISBN 978-0-12-643732-4

- Emsley, J. (1980). "Very Strong Hydrogen Bonds". Chemical Society Reviews. 9 (1): 91–124. doi:10.1039/cs9800900091

- Latimer, Wendell M.; Rodebush, Worth H. (1920). "Polarity and ionization from the standpoint of the Lewis theory of valence.". Journal of the American Chemical Society. 42 (7): 1419–1433. doi:10.1021/ja01452a015

- Campbell, Neil A.; Brad Williamson; Robin J. Heyden (2006). Biology: Exploring Life. Boston, Massachusetts: Pearson Prentice Hall. ISBN 0-13-250882-6

- Markovitch, Omer; Agmon, Noam (2007). "Structure and energetics of the hydronium hydration shells". J. Phys. Chem. A. 111 (12): 2253–2256. PMID 17388314. doi:10.1021/jp068960g.

- Gilman-Politi, Regina; Harries, Daniel (2011). "Unraveling the Molecular Mechanism of Enthalpy Driven Peptide Folding by Polyol Osmolytes". Journal of Chemical Theory and Computation. 7 (11): 3816–3828. PMID 26598272. doi:10.1021/ct200455n

- Pauling, L. (1960). The nature of the chemical bond and the structure of molecules and crystals; an introduction to modern structural chemistry (3rd ed.). Ithaca (NY): Cornell University Press. p. 450. ISBN 0-8014-0333-2

- Jorgensen, W. L.; Madura, J. D. (1985). "Temperature and size dependence for Monte Carlo simulations of TIP4P water". Mol. Phys. 56 (6): 1381. Bibcode:1985MolPh..56.1381J. doi:10.1080/00268978500103111

- Ghanty, Tapan K.; Staroverov, Viktor N.; Koren, Patrick R.; Davidson, Ernest R. (2000-02-01). "Is the Hydrogen Bond in Water Dimer and Ice Covalent?". Journal of the American Chemical Society. 122 (6): 1210–1214. ISSN 0002-7863. doi:10.1021/ja9937019

Chemistry of Cell and Living Organisms

Cell is the basic unit of all living organisms. It consists of a membrane with a cytoplasm within it. Cells form tissues and tissues collectively form organs. Cells are mainly of two types, prokaryotic and eukaryotic. The aspects elucidated in this chapter are of vital importance, and provide a better understanding of living cells.

Cell

A cell is a microscopic, structural and functional unit of all living organisms capable of independent existence. Some functioning cells come together to form a tissue and tissues collectively form organs. In more complex living organisms, organs work together for the purpose of survival as system. However, in all living organisms, the cell is a functional unit and all of biology revolves around the activity of the cell.

The word 'cell' was first coined by Robert Hooke in 1665 to designate the empty honey-comb like structures viewed in a thin section of bottle cork, which he examined. In 1838, the German botanist Matthias Schleiden proposed that all the plants are made up of plant cells. Then in 1839, his colleague, the anatomist Theodore Schwann studied and concluded that all animals are also composed of animal cells. But still the real nature of a cell was in doubt. Cell theory was again rewritten by Rudolf Virchow in 1858.

In his theory he said that all living things are made up of cells and that all cells arise from pre-existing cells. It was German biologist Schulze who found in 1861 that the cells are not empty as were seen by Hooke but contain a 'stuff' of life called protoplasm.

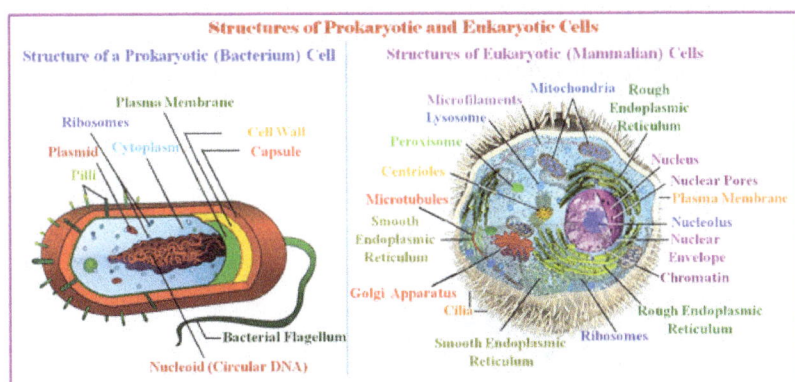

The structure of Prokaryotic and Eukaryotic Cells.

During the 1950s scientists developed the concept that all organisms may be classified as prokaryotes or eukaryotes. For example, in prokaryotic cells, there is no nucleus; eukaryotic cells have a nucleus. Another important difference between prokaryotes and eukaryotes is that the prokaryotic cell does not have any intracellular components. Bacteria and blue- green algae come under the prokaryotic group, and protozoa, fungi, animals, and plants come under the eukaryotic group.

Modern Cell Theory

Modern biologists have made certain additions to the original cell theory, which now states that:

- All organisms are made up of cells.

- New cells are always produced from pre-existing cells.

- The cell is a structural and functional unit of all living things.

- The cell contains hereditary information which is passed on from cell to cell during cell division.

- All cells are basically the same in chemical composition and metabolic activities.

Living Organisms

So we knew that all living things are composed of one or more cells and the products of those cells. The chemical compounds that make up the structures in cells are a mixture of organic compounds and inorganic compounds. Organic compounds always contain carbon and hydrogen (and maybe some other elements), inorganic compounds do not contain carbon and hydrogen together. Simply, Organic refers to life and inorganic compounds make up non-living substances. Organic compounds are found in living things, their wastes, and their remains. Examples of organic compounds that are basic to life includes: carbohydrates (sugars, starches), lipids (fats & waxes), proteins, enzymes, nucleic acids (DNA & RNA). Examples of inorganic compounds includes: water, carbon dioxide. The elements (atoms) in organic compounds are held together by covalent bonds, which form as a result of the sharing of two electrons between two atoms.

The Composition of Living Cell

All living organism contains the following organic molecules for their lives:

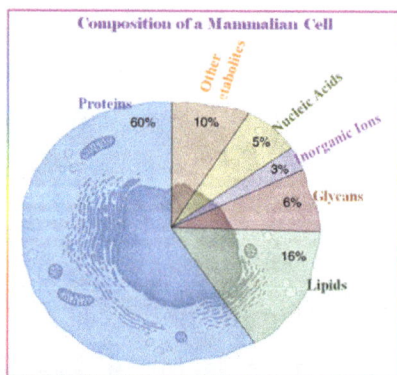

The composition of mammalian Cell

Carbohydrates as One of the Major Components of Living Organism:

- Main source of energy for living organisms: sugar and starch.

- Sugar is broken down inside the body into glucose, which is used for energy.

- Excess sugar is stored as starch.

Elements Present	Used by organisms for ...	Building Block	Related Saccharides
carbon hydrogen oxygen H:O = 2:1 always	energy structure	monosaccharides (simple sugars) ex: glucose	**dissaccharide** 2 connected monosaccarides (ex: maltose) **polysaccharide** 3 or more connected monosaccarides (ex: starch, glycogen, chitin, cellulose)

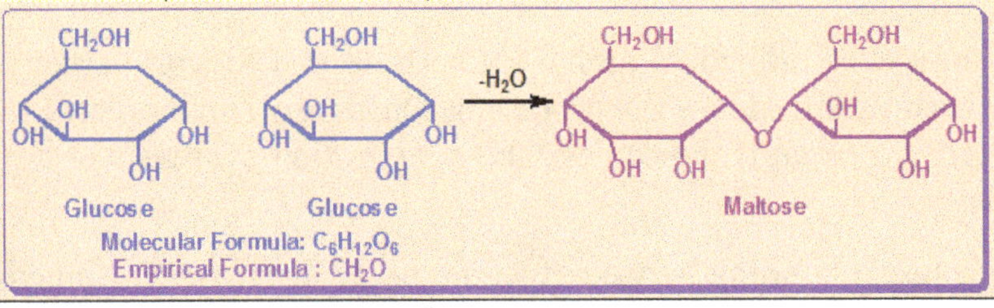

Glucose + Glucose $\xrightarrow{-H_2O}$ Maltose

Molecular Formula: $C_6H_{12}O_6$
Empirical Formula : CH_2O

- Chitin and cellulose are examples of carbohydrates with structural functions. Chitin is the material that makes up the exoskeletons of all arthropods (insects, spiders, lobsters, etc.). Cellulose is what the cell wall in plant cells is made of.

- Starch is the form by which plants store extra carbohydrates. Glycogen, sometimes referred to as animal starch, is the form by which animals store extra carbohydrates. We store glycogen in our livers.

Proteins are the Major Components of Living Organism

- amino acids are the building blocks of proteins

- growth and repair

- build body parts

- provide energy

- carry oxygen in blood

- fight germs

- make hormones

- Enzymes: special type of protein that regulates chemical activities in the body.

- There is an "N" in the word proteiN. The element nitrogen is always present in proteiNs.

Elements Present	Used by organisms for ...	Related Terms & Info
carbon hydrogen oxygen **NITROGEN** (always those 4) phosphorus sulfur (possibly)	structure & movement (muscles) enzymes antibodies hormones pigments	peptide bond = the bond that holds amino acids together in protein molecules dipeptide = two connected amino acids polypeptide = 3 or more connected amino acids
Building Block of Proteins:	General Structure of Amino Acids	

General Structure of Amino Acids

- By what process are individual amino acids combined to from larger proteiNs? It is the dehydration synthesis. This is the process by which any small organic molecules are combined to form big organic molecules. The dehydration synthesis of a protein is typically illustrated as:

- Dipeptide is just a bonding between two amino acids. If we continued to add more and more amino acids to the dipeptide we would then call the molecule a Polypeptide and then we would end up with large protein molecules.

- Dipeptide, polypeptide, peptide bonds are all protein stuff.

- The hydrolysis (breakdown) of a dipeptide/polypeptide can be summarized as:

Dipeptide / Polypeptide + Water → Amino acids + Amino acids

- On hydrolysis, water is added at the beginning and the products are smaller than the molecule we start with.

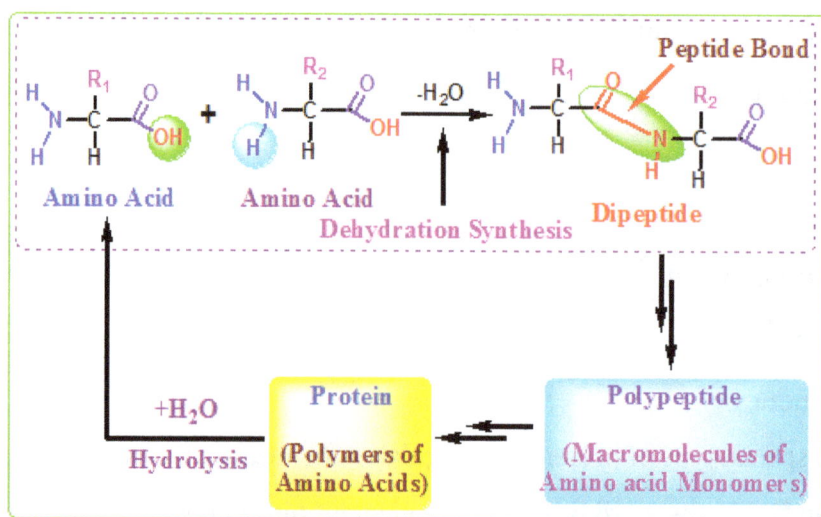

Lipid

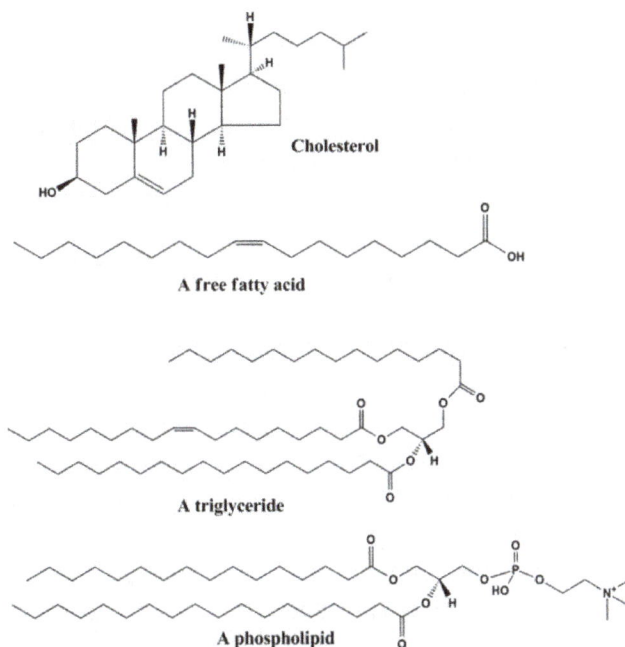

Structures of some common lipids. At the top are cholesterol and oleic acid. The middle structure is a triglyceride composed of oleoyl, stearoyl, and palmitoyl chains attached to a glycerol backbone. At the bottom is the common phospholipid phosphatidylcholine.

In biology, lipid is a loosely defined term for substances of biological origin that are soluble in nonpolar solvents. It comprises a group of naturally occurring molecules that include fats, waxes, sterols, fat-soluble vitamins (such as vitamins A, D, E, and K), monoglycerides, diglycerides, triglycerides, phospholipids, and others. The main biological functions of lipids include storing energy, signaling, and acting as structural components of cell membranes. Lipids have applications in the cosmetic and food industries as well as in nanotechnology.

Scientists may broadly define lipids as hydrophobic or amphiphilic small molecules; the amphiphilic nature of some lipids allows them to form structures such as vesicles, multilamellar/unilamellar liposomes, or membranes in an aqueous environment. Biological lipids originate entirely or in part from two distinct types of biochemical subunits or "building-blocks": ketoacyl and isoprene groups. Using this approach, lipids may be divided into eight categories: fatty acids, glycerolipids, glycerophospholipids, sphingolipids, saccharolipids, and polyketides (derived from condensation of ketoacyl subunits); and sterol lipids and prenol lipids (derived from condensation of isoprene subunits).

Although the term *lipid* is sometimes used as a synonym for fats, fats are a subgroup of lipids called triglycerides. Lipids also encompass molecules such as fatty acids and their derivatives (including tri-, di-, monoglycerides, and phospholipids), as well as other sterol-containing metabolites such as cholesterol. Although humans and other mammals use various biosynthetic pathways both to break down and to synthesize lipids, some essential lipids cannot be made this way and must be obtained from the diet.

The word *lipid* stems etymologically from the Greek *lipos* (fat).

Categories of Lipids

Fatty Acids

Fatty acids, or fatty acid residues when they are part of a lipid, are a diverse group of molecules synthesized by chain-elongation of an acetyl-CoA primer with malonyl-CoA or methylmalonyl-CoA groups in a process called fatty acid synthesis. They are made of a hydrocarbon chain that terminates with a carboxylic acid group; this arrangement confers the molecule with a polar, hydrophilic end, and a nonpolar, hydrophobic end that is insoluble in water. The fatty acid structure is one of the most fundamental categories of biological lipids, and is commonly used as a building-block of more structurally complex lipids. The carbon chain, typically between four and 24 carbons long, may be saturated or unsaturated, and may be attached to functional groups containing oxygen, halogens, nitrogen, and sulfur. If a fatty acid contains a double bond, there is the possibility of either a *cis* or *trans* geometric isomerism, which significantly affects the molecule's configuration. *Cis*-double bonds cause the fatty acid chain to bend, an effect that is compounded with more double bonds in the chain. Three double bonds in 18-carbon *linolenic acid*, the most abundant fatty-acyl chains of plant *thylakoid membranes*, render these membranes highly *fluid* despite environmental low-temperatures, and also makes linolenic acid give dominating sharp peaks in high resolution 13-C NMR spectra of chloroplasts. This in turn plays an important role in the structure and function of cell membranes. Most naturally occurring fatty acids are of the *cis* configuration, although the *trans* form does exist in some natural and partially hydrogenated fats and oils.

Examples of biologically important fatty acids include the eicosanoids, derived primarily from arachidonic acid and eicosapentaenoic acid, that include prostaglandins, leukotrienes, and thromboxanes. Docosahexaenoic acid is also important in biological systems, particularly with respect to sight. Other major lipid classes in the fatty acid category are the fatty esters and fatty amides. Fatty esters include important biochemical intermediates such as wax esters, fatty acid thioester coenzyme A derivatives, fatty acid thioester ACP derivatives and fatty acid carnitines. The fatty amides include N-acyl ethanolamines, such as the cannabinoid neurotransmitter anandamide.

Glycerolipids

Glycerolipids are composed of mono-, di-, and tri-substituted glycerols, the best-known being the fatty acid triesters of glycerol, called triglycerides. The word "triacylglycerol" is sometimes used synonymously with "triglyceride". In these compounds, the three hydroxyl groups of glycerol are each esterified, typically by different fatty acids. Because they function as an energy store, these lipids comprise the bulk of storage fat in animal tissues. The hydrolysis of the ester bonds of triglycerides and the release of glycerol and fatty acids from adipose tissue are the initial steps in metabolizing fat.

Additional subclasses of glycerolipids are represented by glycosylglycerols, which are characterized by the presence of one or more sugar residues attached to glycerol via a glycosidic linkage. Examples of structures in this category are the digalactosyldiacylglycerols found in plant membranes and seminolipid from mammalian sperm cells.

Glycerophospholipids

Glycerophospholipids, usually referred to as phospholipids, are ubiquitous in nature and are key components of the lipid bilayer of cells, as well as being involved in metabolism and cell signaling. Neural tissue (including the brain) contains relatively high amounts of glycerophospholipids, and alterations in their composition has been implicated in various neurological disorders. Glycerophospholipids may be subdivided into distinct classes, based on the nature of the polar headgroup at the *sn*-3 position of the glycerol backbone in eukaryotes and eubacteria, or the *sn*-1 position in the case of archaebacteria.

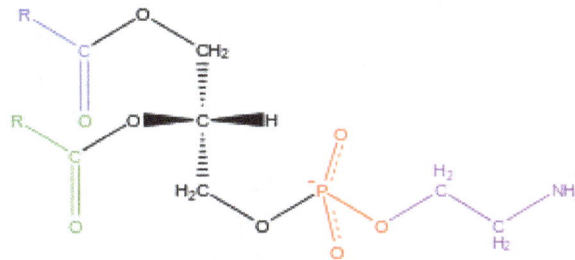

Phosphatidylethanolamine

Examples of glycerophospholipids found in biological membranes are phosphatidylcholine (also known as PC, GPCho or lecithin), phosphatidylethanolamine (PE or GPEtn) and phosphatidylserine (PS or GPSer). In addition to serving as a primary component of cellular membranes and binding sites for intra- and intercellular proteins, some glycerophospholipids in eukaryotic cells, such as phosphatidylinositols and phosphatidic acids are either precursors of or, themselves, membrane-derived second messengers. Typically, one or both of these hydroxyl groups are acylated with long-chain fatty acids, but there are also alkyl-linked and 1Z-alkenyl-linked (plasmalogen) glycerophospholipids, as well as dialkylether variants in archaebacteria.

Sphingolipids

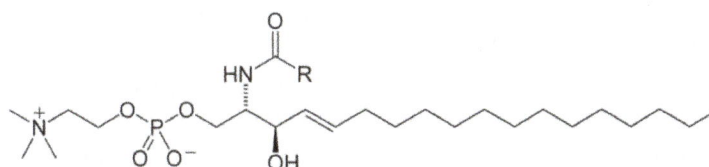

Sphingomyelin

Sphingolipids are a complicated family of compounds that share a common structural feature, a sphingoid base backbone that is synthesized *de novo* from the amino acid serine and a long-chain fatty acyl CoA, then converted into ceramides, phosphosphingolipids, glycosphingolipids and other compounds. The major sphingoid base of mammals is commonly referred to as sphingosine. Ceramides (N-acyl-sphingoid bases) are a major subclass of sphingoid base derivatives with an amide-linked fatty acid. The fatty acids are typically saturated or mono-unsaturated with chain lengths from 16 to 26 carbon atoms.

The major phosphosphingolipids of mammals are sphingomyelins (ceramide phosphocholines), whereas insects contain mainly ceramide phosphoethanolamines and fungi have phytoceramide phosphoinositols and mannose-containing headgroups. The glycosphingolipids are a diverse family of molecules composed of one or more sugar residues linked via a glycosidic bond to the sphingoid base. Examples of these are the simple and complex glycosphingolipids such as cerebrosides and gangliosides.

Sterol Lipids

Sterol lipids, such as cholesterol and its derivatives, are an important component of membrane lipids, along with the glycerophospholipids and sphingomyelins. The steroids, all derived from the same fused four-ring core structure, have different biological roles as hormones and signaling molecules. The eighteen-carbon (C18) steroids include the estrogen family whereas the C19 steroids comprise the androgens such as testosterone and androsterone. The C21 subclass includes the progestogens as well as the glucocorticoids and mineralocorticoids. The secosteroids, comprising various forms of vitamin D, are characterized by cleavage of the B ring of the core structure. Other examples of sterols are the bile acids and their conjugates, which in mammals are oxidized derivatives of cholesterol and are synthesized in the liver. The plant equivalents are the phytosterols, such as β-sitosterol, stigmasterol, and brassicasterol; the latter compound is also used as a biomarker for algal growth. The predominant sterol in fungal cell membranes is ergosterol.

Prenol Lipids

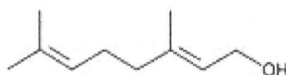

Prenol lipid (2E-geraniol)

Prenol lipids are synthesized from the five-carbon-unit precursors isopentenyl diphosphate and dimethylallyl diphosphate that are produced mainly via the mevalonic acid (MVA) pathway. The simple isoprenoids (linear alcohols, diphosphates, etc.) are formed by the successive addition of C5 units, and are classified according to number of these terpene units. Structures containing greater than 40 carbons are known as polyterpenes. Carotenoids are important simple isoprenoids that function as antioxidants and as precursors of vitamin A. Another biologically important class of molecules is exemplified by the quinones and hydroquinones, which contain an isoprenoid tail attached to a quinonoid core of non-isoprenoid origin. Vitamin E and vitamin K, as well as the ubiquinones, are examples of this class. Prokaryotes synthesize polyprenols (called bactoprenols) in which the terminal isoprenoid unit attached to oxygen remains unsaturated, whereas in animal polyprenols (dolichols) the terminal isoprenoid is reduced.

Saccharolipids

Saccharolipids describe compounds in which fatty acids are linked directly to a sugar backbone, forming structures that are compatible with membrane bilayers. In the saccharolipids, a monosaccharide substitutes for the glycerol backbone present in glycerolipids and glycerophospholipids. The most familiar saccharolipids are the acylated glucosamine precursors of the Lipid A component of the lipopolysaccharides in Gram-negative bacteria. Typical lipid A molecules are disaccharides of glucosamine, which are derivatized with as many as seven fatty-acyl chains. The minimal lipopolysaccharide required for growth in *E. coli* is Kdo_2-Lipid A, a hexa-acylated disaccharide of glucosamine that is glycosylated with two 3-deoxy-D-man-no-octulosonic acid (Kdo) residues.

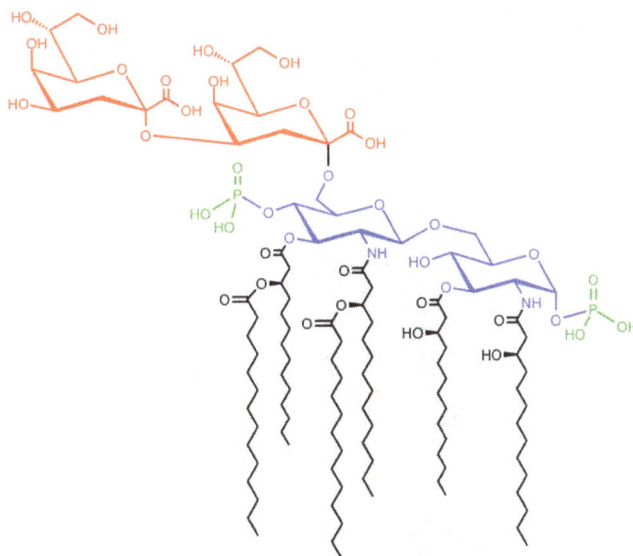

Structure of the saccharolipid Kdo_2-lipid A. Glucosamine residues in blue, Kdo residues in red, acyl chains in black and phosphate groups in green.

Polyketides

Polyketides are synthesized by polymerization of acetyl and propionyl subunits by classic enzymes as well as iterative and multimodular enzymes that share mechanistic features with the fatty acid synthases. They comprise a large number of secondary metabolites and natural products from animal, plant, bacterial, fungal and marine sources, and have great structural diversity. Many polyketides are cyclic molecules whose backbones are often further modified by glycosylation, methylation, hydroxylation, oxidation, and/or other processes. Many commonly used anti-microbial, anti-parasitic, and anti-cancer agents are polyketides or polyketide derivatives, such as erythromycins, tetracyclines, avermectins, and antitumor epothilones.

Biological Functions

Membranes

Eukaryotic cells feature compartmentalized membrane-bound organelles that carry out different biological functions. The glycerophospholipids are the main structural component of biological membranes, such as the cellular plasma membrane and the intracellular membranes of organelles; in animal cells the plasma membrane physically separates the intracellular components from the extracellular environment. The glycerophospholipids are amphipathic molecules (containing both hydrophobic and hydrophilic regions) that contain a glycerol core linked to two fatty acid-derived "tails" by ester linkages and to one "head" group by a phosphate ester linkage. While glycerophospholipids are the major component of biological membranes, other non-glyceride lipid components such as sphingomyelin and sterols (mainly cholesterol in animal cell membranes) are also found in biological membranes. In plants and algae, the galactosyldiacylglycerols, and sulfoquinovosyldiacylglycerol, which lack a phosphate group, are important components of membranes of chloroplasts and related organelles and are the most abundant lipids in photosynthetic tissues, including those of higher plants, algae and certain bacteria.

Plant thylakoid membranes have the largest lipid component of a non-bilayer forming monogalactosyl diglyceride (MGDG), and little phospholipids; despite this unique lipid composition, chloroplast thylakoid membranes have been shown to contain a dynamic lipid-bilayer matrix as revealed by magnetic resonance and electron microscope studies.

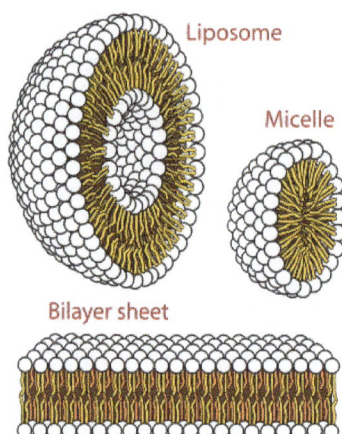

Self-organization of phospholipids: a spherical liposome, a micelle, and a lipid bilayer.

A biological membrane is a form of lamellar phase lipid bilayer. The formation of lipid bilayers is an energetically preferred process when the glycerophospholipids described above are in an aqueous environment. This is known as the hydrophobic effect. In an aqueous system, the polar heads of lipids align towards the polar, aqueous environment, while the hydrophobic tails minimize their contact with water and tend to cluster together, forming a vesicle; depending on the concentration of the lipid, this biophysical interaction may result in the formation of micelles, liposomes, or lipid bilayers. Other aggregations are also observed and form part of the polymorphism of amphiphile (lipid) behavior. Phase behavior is an area of study within biophysics and is the subject of current academic research. Micelles and bilayers form in the polar medium by a process known as the hydrophobic effect. When dissolving a lipophilic or amphiphilic substance in a polar environment, the polar molecules (i.e., water in an aqueous solution) become more ordered around the dissolved lipophilic substance, since the polar molecules cannot form hydrogen bonds to the lipophilic areas of the amphiphile. So in an aqueous environment, the water molecules form an ordered "clathrate" cage around the dissolved lipophilic molecule.

The formation of lipids into protocell membranes represents a key step in models of abiogenesis, the origin of life.

Energy Storage

Triglycerides, stored in adipose tissue, are a major form of energy storage both in animals and plants. The adipocyte, or fat cell, is designed for continuous synthesis and breakdown of triglycerides in animals, with breakdown controlled mainly by the activation of hormone-sensitive enzyme lipase. The complete oxidation of fatty acids provides high caloric content, about 9 kcal/g, compared with 4 kcal/g for the breakdown of carbohydrates and proteins. Migratory birds that must fly long distances without eating use stored energy of triglycerides to fuel their flights.

Signaling

In recent years, evidence has emerged showing that lipid signaling is a vital part of the cell signaling. Lipid signaling may occur via activation of G protein-coupled or nuclear receptors, and members of several different lipid categories have been identified as signaling molecules and cellular messengers. These include sphingosine-1-phosphate, a sphingolipid derived from ceramide that is a potent messenger molecule involved in regulating calcium mobilization, cell growth, and apoptosis; diacylglycerol (DAG) and the phosphatidylinositol phosphates (PIPs), involved in calcium-mediated activation of protein kinase C; the prostaglandins, which are one type of fatty-acid derived eicosanoid involved in inflammation and immunity; the steroid hormones such as estrogen, testosterone and cortisol, which modulate a host of functions such as reproduction, metabolism and blood pressure; and the oxysterols such as 25-hydroxy-cholesterol that are liver X receptor agonists. Phosphatidylserine lipids are known to be involved in signaling for the phagocytosis of apoptotic cells and/or pieces of cells. They accomplish this by being exposed to the extracellular face of the cell membrane after the inactivation of flippases which place them exclusively on the cytosolic side and the activation of scramblases, which scramble the orientation of the phospholipids. After this occurs, other cells recognize the phosphatidylserines and phagocytosize the cells or cell fragments exposing them.

Other Functions

The "fat-soluble" vitamins (A, D, E and K) – which are isoprene-based lipids – are essential nutrients stored in the liver and fatty tissues, with a diverse range of functions. Acyl-carnitines are involved in the transport and metabolism of fatty acids in and out of mitochondria, where they undergo beta oxidation. Polyprenols and their phosphorylated derivatives also play important transport roles, in this case the transport of oligosaccharides across membranes. Polyprenol phosphate sugars and polyprenol diphosphate sugars function in extra-cytoplasmic glycosylation reactions, in extracellular polysaccharide biosynthesis (for instance, peptidoglycan polymerization in bacteria), and in eukaryotic protein N-glycosylation. Cardiolipins are a subclass of glycerophospholipids containing four acyl chains and three glycerol groups that are particularly abundant in the inner mitochondrial membrane. They are believed to activate enzymes involved with oxidative phosphorylation. Lipids also form the basis of steroid hormones.

Metabolism

The major dietary lipids for humans and other animals are animal and plant triglycerides, sterols, and membrane phospholipids. The process of lipid metabolism synthesizes and degrades the lipid stores and produces the structural and functional lipids characteristic of individual tissues.

Biosynthesis

In animals, when there is an oversupply of dietary carbohydrate, the excess carbohydrate is converted to triglycerides. This involves the synthesis of fatty acids from acetyl-CoA and the esterification of fatty acids in the production of triglycerides, a process called lipogenesis. Fatty acids are made by fatty acid synthases that polymerize and then reduce acetyl-CoA units. The acyl chains in the fatty acids are extended by a cycle of reactions that add the acetyl group, reduce it to an alcohol, dehydrate it to an alkene group and then reduce it again to an alkane group. The enzymes of fatty acid biosynthesis are divided into two groups, in animals and fungi all these fatty acid synthase reactions are carried out by a single multifunctional protein, while in plant plastids and bacteria separate enzymes perform each step in the pathway. The fatty acids may be subsequently converted to triglycerides that are packaged in lipoproteins and secreted from the liver.

The synthesis of unsaturated fatty acids involves a desaturation reaction, whereby a double bond is introduced into the fatty acyl chain. For example, in humans, the desaturation of stearic acid by stearoyl-CoA desaturase-1 produces oleic acid. The doubly unsaturated fatty acid linoleic acid as well as the triply unsaturated α-linolenic acid cannot be synthesized in mammalian tissues, and are therefore essential fatty acids and must be obtained from the diet.

Triglyceride synthesis takes place in the endoplasmic reticulum by metabolic pathways in which acyl groups in fatty acyl-CoAs are transferred to the hydroxyl groups of glycerol-3-phosphate and diacylglycerol.

Terpenes and isoprenoids, including the carotenoids, are made by the assembly and modification of isoprene units donated from the reactive precursors isopentenyl pyrophosphate and dimethylallyl pyrophosphate. These precursors can be made in different ways. In animals and archaea, the mevalonate pathway produces these compounds from acetyl-CoA, while in plants and bacteria the non-mevalonate pathway uses pyruvate and glyceraldehyde 3-phosphate as substrates. One important reaction that uses these activated isoprene donors is steroid biosynthesis. Here, the isoprene units are joined together to make squalene and then folded up and formed into a set of rings to make lanosterol. Lanosterol can then be converted into other steroids such as cholesterol and ergosterol.

Degradation

Beta oxidation is the metabolic process by which fatty acids are broken down in the mitochondria and/or in peroxisomes to generate acetyl-CoA. For the most part, fatty acids are oxidized by a mechanism that is similar to, but not identical with, a reversal of the process of fatty acid synthesis. That is, two-carbon fragments are removed sequentially from the carboxyl end of the acid after steps of dehydrogenation, hydration, and oxidation to form a beta-keto acid, which is split by thiolysis. The acetyl-CoA is then ultimately converted into ATP, CO_2, and H_2O using the citric acid cycle and the electron transport chain. Hence the citric acid cycle can start at acetyl-CoA when fat is being broken down for energy if there is little or no glucose available. The energy yield of the complete oxidation of the fatty acid palmitate is 106 ATP. Unsaturated and odd-chain fatty acids require additional enzymatic steps for degradation.

Nutrition and Health

Most of the fat found in food is in the form of triglycerides, cholesterol, and phospholipids. Some dietary fat is necessary to facilitate absorption of fat-soluble vitamins (A, D, E, and K) and carotenoids. Humans and other mammals have a dietary requirement for certain essential fatty acids, such as linoleic acid (an omega-6 fatty acid) and alpha-linolenic acid (an omega-3 fatty acid) because they cannot be synthesized from simple precursors in the diet. Both of these fatty acids are 18-carbon polyunsaturated fatty acids differing in the number and position of the double bonds. Most vegetable oils are rich in linoleic acid (safflower, sunflower, and corn oils). Alpha-linolenic acid is found in the green leaves of plants, and in selected seeds, nuts, and legumes (in particular flax, rapeseed, walnut, and soy). Fish oils are particularly rich in the longer-chain omega-3 fatty acids eicosapentaenoic acid (EPA) and docosahexaenoic acid (DHA). A large number of studies have shown positive health benefits associated with consumption of omega-3 fatty acids on infant development, cancer, cardio-vascular diseases, and various mental illnesses, such as depression, attention-deficit hyperactivity disorder, and dementia. In contrast, it is now well-established that consumption of trans fats, such as those present in partially hydrogenated vegetable oils, are a risk factor for cardiovascular disease.

A few studies have suggested that total dietary fat intake is linked to an increased risk of obesity and diabetes. However, a number of very large studies, including the Women's Health Initiative Dietary Modification Trial, an eight-year study of 49,000 women, the Nurses' Health Study and the Health Professionals Follow-up Study, revealed no such links. None of these studies suggested any connection between percentage of calories from fat and risk of cancer, heart disease, or weight gain. The Nutrition Source, a website maintained by the Department of Nutrition at the Harvard School of Public Health, summarizes the current evidence on the impact of dietary fat: "Detailed research—much of it done at Harvard—shows that the total amount of fat in the diet isn't really linked with weight or disease."

Lipids (Fats, Oils, and Waxes) are Another Component of Living Organisms:

- Lipids are third group of organic compounds present in cell of all living organisms. Lipids contain C, H, and O, and that's it. No other elements in lipid molecules are present.

- Carbohydrates always have twice as many hydrogen atoms as oxygen atoms (H:O ratio = 2:1). Lipids never do. Also, the structural formulas of carbohydrates have the ring while lipids do not.

- Lipids are energy rich compounds.

- A fatty acid is nothing more than a long C-H chain with a carboxyl group (COOH) on the end.

- The carboxyl group gives a molecule an acidic property. Both of the organic acids fatty Acids and amino Acids have carboxyl groups.

- Glycerol is classified as an alcohol (due to the OH's). It always looks the same: Three C's with Three -OH's and everything else H's.

Elements Present	Used by Organisms for ...	Related Terms & Info
Carbon Hydrogen Oxygen ONLY ! There is no specific H:O ratio.	Stored Energy Structure (important part of cell membranes)	saturated fat = C-C bonds are all single bonds unsaturated fat = contain at least one double or triple C-C bond
Building Blocks of Lipids	Fatty Acid	Glycerol

- Combining three fatty acids with one glycerol by the process of DEHYDRATION SYNTHESIS give fatty acids.

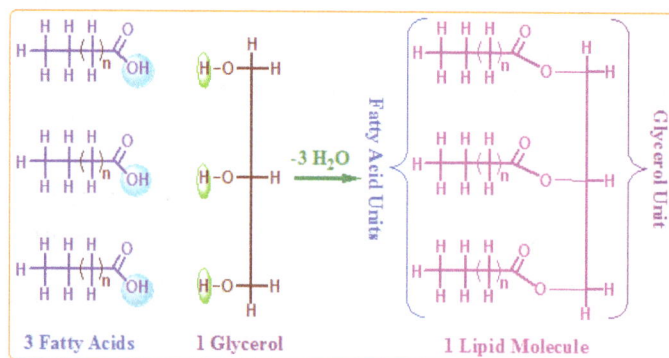

3 Fatty Acids   1 Glycerol   1 Lipid Molecule

- There is no Nitrogen anywhere, so this is definitely not a proteiN reaction.

- Also there are no ring-shaped molecules, so one is not dealing with carbohydrates.

- The hydrolysis (digestion) of a lipid is similar in living organism as is the case of Carbohydrate/proteins and can be summarized as below:

Lipid + Water → 3 Fatty Acids + 1 Glycerol

Nucleic Acids-DNA and RNA: 5% of the Total Components in Mammalian Cell

- "Blueprints" of life

- Store information that the body needs to build proteins

- DNA (deoxyribonucleic acid)

- Stores information; delivers.

- DNA & RNA (like proteins, carbohydrates, and lipids) are polymers--- long chains of smaller repeating units. The repeating unit in nucleic acids is called a Nucleotide.

- Every nucleotide has the same basic structure as below:

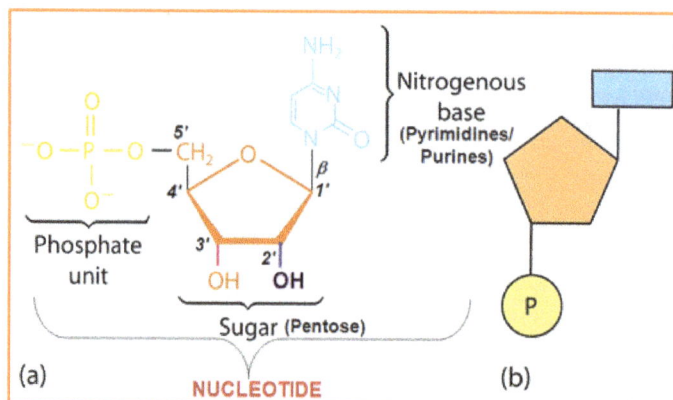

- Comparison of DNA and RNA:

	DNA	RNA
Full Name	Deoxyribonucleic acid	Ribonucleic acid
Basic Structure	Two long twisting strands of nucleotides in the form of a "double helix"	One single strand of nucleotides
Nucleotide Sugar	2'-Deoxyribose	Ribose
Nitrogenous Bases	Guanine (G) Cytosine (C) Adenine (A) Thymine (T)	Guanine (G) Cytosine (C) Adenine (A) Uracil (U)
Location in a Cell	Nucleus (the Chromosomes)	Nucleus, in the Cytoplasm, and at the Ribosomes
Function	The Hereditary Material of a Cell, Directs and Controls Cell Activities	Involved in Protein Synthesis

- So, DNA & RNA are alike in that they are both nucleic acids composed of nucleotides.

- Their differences lie in their functions and structure.

- The main structural differences are the number of strands in the molecule, the sugar structure, and one of the N-bases (thymine in DNA, uracil in RNA).

Chemical Reaction in Living Cell

Therefore, these four major types of macromolecules found in living cells—carbohydrates, lipids, proteins, and nucleic acids--are made of small, repeating subunits called monomers. The monomers are not always identical but they always have similar chemical structures. They are joined together by a series of chemical bond formed via the reactions called polymerisation to form large, complex molecules called polymers.

The Four Major Types of Macromolecules Found in Living Cells				
Macromolecule	Elements	Monomer	Polymer	example
Carbohydrate	C, H, O	Simple sugars	Polysaccharide	Starch
Lipids	C, H, O	Fatty acids and glycerol	Lipid	Fats, oils, waxes
Proteins	C, H, O, N, S	Amino acids	Polypeptides	Insulin
Nucleic acids	C, H, O, P	Nucleotides	Nucleic acids	DNA

Macromolecular functions are directly related to their structures, shapes and to the chemical properties which is similar to their monomers. The way the monomers are arranged in the macromolecule determines its shape and function in the similar way that the arrangement of the letters in a word determine its sound and meaning.

Much of a cell's activities involve the proper organization and bonding of macromolecules and their inter/intra-molecular interactions with other macromolecules. It is the job of DNA both directly and indirectly to coordinate and direct these activities. An understanding of the structure and functions of carbohydrates and lipids is not particularly key to the understanding of molecular family; however, they play a crucial role in maintaining the cell structure and functions.

Chemical Reaction in Living Cell: Dehydration Synthesis vs. Hydrolysis

The chemical process that connects the smaller subunits to form large organic macromolecules is called dehydration synthesis. Hydrolysis is the process that breaks large organic macromolecules into their smaller subunits. It is the opposite of dehydration synthesis. In hydrolysis, water is added and the large compounds are split into small fragments. In living system, the process of hydrolysis is involved in digestion --- when food is broken down into nutrients.

Process	Start with ...	Ends with ...	Example
Dehydration Synthesis	small molecules (subunits/monomers)	large molecules and water (macromolecules)	
Hydrolysis	water and large molecules (macromolecules)	small molecules (subunits/monomers)	Digestion

How to Visualize Biomolecules in Living Cells?

As was discussed earlier, Living systems are composed of networks of several interacting biopolymers, ions and metabolites. These cellular components drive a complex array of cellular processes, many of which cannot be observed when the biomolecules are examined in their purified, isolated forms. Therefore, researchers have begun to study biological processes in living cells and in whole organisms instead of testing in laboratory in test tubes. To do so tracking the molecules is necessary within the cell's native environments. Direct detection of few biomolecules in complex biomolecular environment is possible but for all other cases we have to depend on indirect detection techniques. Thus, several methods have been developed to equip cellular components with reporter tags for visualization and isolation from biological samples.

The most popular strategy for cellular imaging involves tagging of the green fluorescent protein (GFP) and its related variants to a biomolecule of interest. Tagging of these fluorescent probes to a target protein enables visualization by fluorescence microscopy. GFP tags can also be used to analyze whole organisms focusing the proteins. Almost every cellular process related to proteins has been studied using GFP like tags.

However, GFP tagging suffers from several short comings such as-(a) tagging causes structural perturbation which in turn influence the protein expression, localization or function; (b) visualized is possible only by optical methods; (c) GFP tagging only be applied to proteneceous materials and cannot be applied to visualize non-proteinaceous components (a significant fraction of cellular biomass) of cell like glycans, lipids, nucleic acids or the thousands of small organic metabolites. Therefore, methods to visualize both proteins, their modifiers, and other non-proteinaceous components would enable us understanding of the whole organism proteome.

To track biomolecules in living cells and whole organisms, Antibody conjugates have been widely used. However, because of the large size and physical properties, access of these reagents to antigens within cells and outside of the vasculature in living animals is a problem.

Therefore, we see that a large molecule tag is not suitable to meet all research need without hampering the cellular activity. Thus, a small molecular fluorescent tagging approach (like tagging of biotin, fluorophores and numerous other small-molecule reporters) has been developed and utilized owing to the availability of reacting centre/functionality within a biomolecule. However, the site-specific chemical modification of biomolecules remains a very difficult task.

Needs for tagging biomolecules uncovered the bioorthogonal chemical reporters strategy to tag biomolecules. Incorporation of unique chemical functionality (a bioorthogonal chemical reporter) into a target biomolecule using the cell's own biosynthetic machinery is the main part of this strategy. Therefore, using this techniques, proteins, glycans and lipids have all been tagged/labeled with chemical reporters in living cells and then ligated with reactive probes. This strategy has also been applied in monitoring enzyme activities and tagging cell surface glycans in whole organisms.

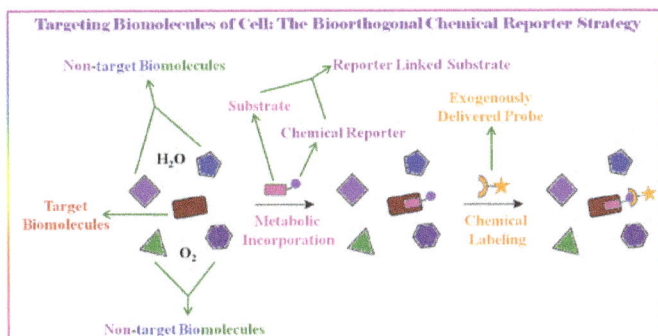

Schematic of bioorthogonal chemical reporters' strategy to visualize cell's biomolecules.

Existing Bioorthogonal Chemical Reporter Systems:

A number of chemical motifs are reported which possess the required qualities of biocompatibility and selective reactivity. Thus they are today well known bioorthogonal chemical reporters in living cells. This group comprises (1) peptide sequences that can be ligated with small-molecule imaging probes, (2) cell surface electrophiles that can be tagged with hydrazide and aminooxy derivatives, (3) azides that can be selectively modified with phosphines or activated alkynes, and (4) terminal alkynes that can be ligated with azides.

Table: Chemical reporters and bioorthogonal reactions used in living systems.

Bioorthogonal Peptide Sequences

The tetracysteine-biarsenical system affords a powerful alternative to GFP tagging for protein visualization.

Ketones and Aldehydes

Ketones, and aldehydes are bioorthogonal chemical reporters that can tag not only proteins, but also glycans and other secondary metabolites.

Azides

The Staudinger Ligation

In contrast to aldehydes and ketones, azides are versatile chemical reporters for labeling all classes of biomolecules in any biological settings. The azides are good electrophiles subject to reaction with soft nucleophiles. This versatile functional group is absent in almost all naturally occurring species. Due to its wonderful bioorthogonality, recently the azide is being used as a chemical reporter in living systems. It is kinetically stable and contains large intrinsic energy. Thus, azides are prone to unique modes of reactivity. Therefore azide has been exploited for the development of bioorthogonal reactions, including the Staudinger ligation of azides with functionalized phosphines and the click reaction {[3+2] cycloaddition} with activated alkynes. These reactions can be used for the selective labeling of azide-functionalized biomolecules.

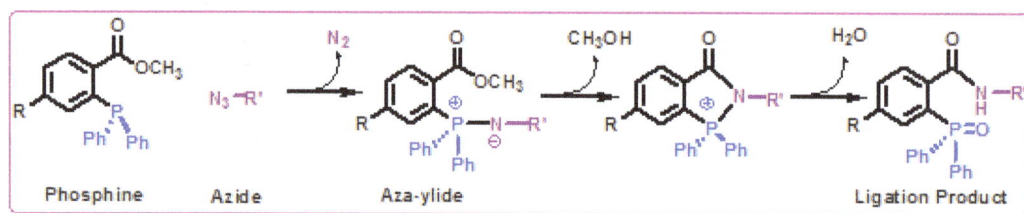

The Staudinger Ligation.

The Staudinger ligation has been used to modify glycans on living cells. Thus, glycoproteins are enriched with ligated components thereby, imparting new functionality to recombinant proteins.

Copper-catalyzed [3+2] azide-alkyne cycloaddition

Azides are also 1,3-dipolar in nature, thus, can undergo reactions with dipolarophiles such as activated alkynes. These π-systems are both extremely rare and inert in biological systems, thus, further increasing the bioorthogonality of the azide along the reaction with alkynes. More than four decades ago, the [3+2] cycloaddition between azides and terminal alkynes to provide stable triazole adducts was first described by Huisgen. The reaction is thermodynamically favorable. Without alkyne activation, however, the process requires stringent reaction conditions (high temperatures or pressures) which are incompatible with living systems. Therefore to make the process facile and thus, compatible with living systems, the alkyne must be activated. One possible way of activating alkynes is to attach an electron withdrawing functional groups like an ester; however, the resulting α,β-unsaturated carbonyl compounds can then act as Michael acceptors for a variety of biological nucleophiles. Therefore this is loosing bioorthogonality.

To make the azide-alkyne cycloaddition a bioorthogonal, one should activate alkyne via activating the terminal alkyne proton by using a catalyst like Cu (I). Thus, the Cu (I)-catalyzed azide-alkyne cycloaddition would be facile at biological temperature and also the rate of the reaction could be faster compared to uncatalysed Staudinger ligation.

Strain-Promoted Cycloaddition

Use of ring strain is an alternative means of activating alkynes for a catalyst-free [3+2] cycloaddition with azides. Constraining the alkyne within an eight-membered ring creates ~18 kcal/mol of strain. This strain energy is released in the transition state upon [3+2] cycloaddition with an azide. As a consequence, cyclooctynes react with azides at room temperature, without the need for a catalyst. Thus, this strain-promoted cycloaddition has been used to label biomolecules both in vitro and on cell surfaces without observable toxic effects. However, the reaction is limited by its slow rate.

How to Introduce Chemical Reporters in Cell's Biomolecules?

In Proteins

To exploit the bioorthogonal chemistry of ketones, azides and alkynes (those functional groups are not present in any natural amino acids) for protein labeling we need using a cell's translational machinery in either a residue-specific or a site-specific manner.

In Glycoconjugates

Azides can be incorporated into glycoconjugates using glycan biosynthetic pathways. Thus, as is shown in figure, azido analog of GlcNAc (GlcNAz) can be incorporated into cytosolic and nuclear glycoproteins.

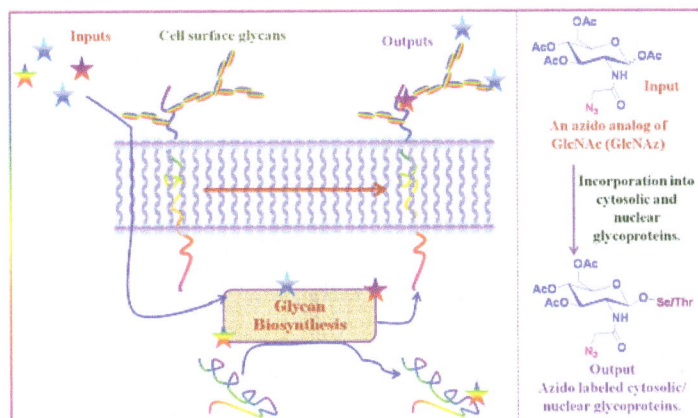

Strategy to incorporate Azides into glycoconjugates.

How to Read Enzyme Function?- Chemical Reporters Can Read

Along with monitoring biomolecule expression and localization, chemical reporters can read enzyme function. Thus, the target protein is labeled with the chemical reporter by virtue of its catalytic activity on a modified substrate. Thus, the activity-based protein profiling approach has been used to monitor enzymatic functions. An alkyne reporter was found to give cleaner labeling than the corresponding azido analog for such purpose.

Example of Bioorthogonal Chemical Reporters in Living Organisms

GFP-protein fusions are widely used for noninvasive imaging of protein expression and localization in living organisms. In a similar manner, both proteins and glycans have been labeled with azides in laboratory mice can be utilized for non invasive imaging.

Encapsulation of Chemistry of Living Cell

As is stated earlier, the bioorthogonal chemical reporter strategy offers a means to visualize many classes of biomolecules in living systems. Substrates linked to chemical reporters can be metabolized by cells and incorporated into proteins, glycans, lipids and other cellular species. After covalent reaction with complementary probes, these classes of biomolecules can be visualized in living cells/living organisms.

Thus, it is clear that chemical reporters and bioorthogonal reactions have a rich future in the field of chemical biology. However, there remains challenge with respect to both metabolic labeling and chemical tagging in biological systems.

References

- Mashaghi S, Jadidi T, Koenderink G, Mashaghi A (February 2013). "Lipid nanotechnology". International Journal of Molecular Sciences. 14 (2): 4242–82. PMC 3588097 . PMID 23429269. doi:10.3390/ijms14024242

- Michelle A, Hopkins J, McLaughlin CW, Johnson S, Warner MQ, LaHart D, Wright JD (1993). Human Biology and Health. Englewood Cliffs, New Jersey, USA: Prentice Hall. ISBN 978-0-13-981176-0

- Hunter JE (November 2006). "Dietary trans fatty acids: review of recent human studies and food industry responses". Lipids. 41 (11): 967–92. PMID 17263298. doi:10.1007/s11745-006-5049-y

- Yashroy RC. (1987). "C NMR studies of lipid fatty acyl chains of chloroplast membranes". Indian Journal of Biochemistry and Biophysics. 24 (6): 177–178

- Brown HA, ed. (2007). Lipodomics and Bioactive Lipids: Mass Spectrometry Based Lipid Analysis. Methods in Enzymology. 423. Boston: Academic Press. ISBN 978-0-12-373895-0

- Hölzl G, Dörmann P (September 2007). "Structure and function of glycoglycerolipids in plants and bacteria". Progress in Lipid Research. 46 (5): 225–43. PMID 17599463. doi:10.1016/j.plipres.2007.05.001

- Yashroy RC. (1990). "Magnetic resonance studies of dynamic organisation of lipids in chloroplast membranes". Journal of Biosciences. 15 (4): 281–288. doi:10.1007/BF02702669

- Hunt SM, Groff JL, Gropper SA (1995). Advanced Nutrition and Human Metabolism. Belmont, California: West Pub. Co. p. 98. ISBN 978-0-314-04467-9

- Wiegandt H (January 1992). "Insect glycolipids". Biochimica et Biophysica Acta. 1123 (2): 117–26. PMID 1739742. doi:10.1016/0005-2760(92)90101-Z

- Grochowski LL, Xu H, White RH (May 2006). "Methanocaldococcus jannaschii uses a modified mevalonate pathway for biosynthesis of isopentenyl diphosphate". Journal of Bacteriology. 188 (9): 3192–8. PMC 1447442 . PMID 16621811. doi:10.1128/JB.188.9.3192-3198.2006

- Deacon J. (2005). Fungal Biology. Cambridge, Massachusetts: Blackwell Publishers. p. 342. ISBN 978-1-4051-3066-0

- Russell DW (2003). "The enzymes, regulation, and genetics of bile acid synthesis". Annual Review of Biochemistry. 72: 137–74. PMID 12543708. doi:10.1146/annurev.biochem.72.121801.161712

- Astrup A (2008). "Dietary management of obesity". JPEN. Journal of Parenteral and Enteral Nutrition. 32 (5): 575–7. PMID 18753397. doi:10.1177/0148607108321707

- Heinz E. (1996). "Plant glycolipids: structure, isolation and analysis", pp. 211–332 in Advances in Lipid Methodology, Vol. 3. W.W. Christie (ed.). Oily Press, Dundee. ISBN 978-0-9514171-6-4

- Malinauskas T (2008). "Docking of fatty acids into the WIF domain of the human Wnt inhibitory factor-1". Lipids. 43 (3): 227–30. PMID 18256869. doi:10.1007/s11745-007-3144-3

Biochemical and Organic Reaction: Analogous Processes

The chemical processes that occur within an organism can be studied under the subject of biochemistry. Organic reactions are chemical reactions which involve organic compounds. This chapter helps the readers in developing a better understanding of the biochemical and organic reaction.

Biochemistry

Biochemistry, sometimes called biological chemistry, is the study of chemical processes within and relating to living organisms. By controlling information flow through biochemical signaling and the flow of chemical energy through metabolism, biochemical processes give rise to the complexity of life. Over the last decades of the 20th century, biochemistry has become so successful at explaining living processes that now almost all areas of the life sciences from botany to medicine to genetics are engaged in biochemical research. Today, the main focus of pure biochemistry is on understanding how biological molecules give rise to the processes that occur within living cells, which in turn relates greatly to the study and understanding of tissues, organs, and whole organisms—that is, all of biology.

Biochemistry is closely related to molecular biology, the study of the molecular mechanisms by which genetic information encoded in DNA is able to result in the processes of life. Depending on the exact definition of the terms used, molecular biology can be thought of as a branch of biochemistry, or biochemistry as a tool with which to investigate and study molecular biology.

Much of biochemistry deals with the structures, functions and interactions of biological macromolecules, such as proteins, nucleic acids, carbohydrates and lipids, which provide the structure of cells and perform many of the functions associated with life. The chemistry of the cell also depends on the reactions of smaller molecules and ions. These can be inorganic, for example water and metal ions, or organic, for example the amino acids, which are used to synthesize proteins. The mechanisms by which cells harness energy from their environment via chemical reactions are known as metabolism. The findings of biochemistry are applied primarily in medicine, nutrition, and agriculture. In medicine, biochemists investigate the causes and cures of diseases. In nutrition, they study how to maintain health and study the effects of nutritional deficiencies. In agriculture, biochemists investigate soil and fertilizers, and try to discover ways to improve crop cultivation, crop storage and pest control.

History

At its broadest definition, biochemistry can be seen as a study of the components and composition of living things and how they come together to become life, and the history of biochemistry may therefore go back as far as the ancient Greeks. However, biochemistry as a specific scientific discipline has its beginning sometime in the 19th century, or a little earlier, depending on which aspect of biochemistry is being focused on. Some argued that the beginning of biochemistry may have been the discovery of the first enzyme, diastase (today called amylase), in 1833 by Anselme Payen, while others considered Eduard Buchner's first demonstration of a complex biochemical process alcoholic fermentation in cell-free extracts in 1897 to be the birth of biochemistry. Some might also point as its beginning to the influential 1842 work by Justus von Liebig, *Animal chemistry, or, Organic chemistry in its applications to physiology and pathology*, which presented a chemical theory of metabolism, or even earlier to the 18th century studies on fermentation and respiration by Antoine Lavoisier. Many other pioneers in the field who helped to uncover the layers of complexity of biochemistry have been proclaimed founders of modern biochemistry, for example Emil Fischer for his work on the chemistry of proteins, and F. Gowland Hopkins on enzymes and the dynamic nature of biochemistry.

Gerty Cori and Carl Cori jointly won the Nobel Prize in 1947 for their discovery of the Cori cycle at RPMI.

The term "biochemistry" itself is derived from a combination of biology and chemistry. In 1877, Felix Hoppe-Seyler used the term (*biochemie* in German) as a synonym for physiological chemistry in the foreword to the first issue of *Zeitschrift für Physiologische Chemie* where he argued for the setting up of institutes dedicated to this field of study. The German chemist Carl Neuberg however is often cited to have coined the word in 1903, while some credited it to Franz Hofmeister.

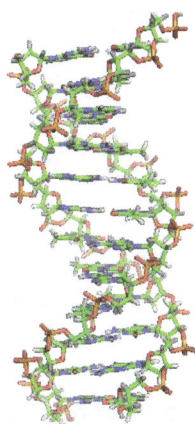

DNA structure (1D65)

It was once generally believed that life and its materials had some essential property or substance (often referred to as the "vital principle") distinct from any found in non-living matter, and it was thought that only living beings could produce the molecules of life. Then, in 1828, Friedrich Wöhler published a paper on the synthesis of urea, proving that organic compounds can be created artificially. Since then, biochemistry has advanced, especially since the mid-20th century, with the development of new techniques such as chromatography, X-ray diffraction, dual polarisation interferometry, NMR spectroscopy, radioisotopic labeling, electron microscopy, and molecular dynamics simulations. These techniques allowed for the discovery and detailed analysis of many molecules and metabolic pathways of the cell, such as glycolysis and the Krebs cycle (citric acid cycle).

Another significant historic event in biochemistry is the discovery of the gene and its role in the transfer of information in the cell. This part of biochemistry is often called molecular biology. In the 1950s, James D. Watson, Francis Crick, Rosalind Franklin, and Maurice Wilkins were instrumental in solving DNA structure and suggesting its relationship with genetic transfer of information. In 1958, George Beadle and Edward Tatum received the Nobel Prize for work in fungi showing that one gene produces one enzyme. In 1988, Colin Pitchfork was the first person convicted of murder with DNA evidence, which led to the growth of forensic science. More recently, Andrew Z. Fire and Craig C. Mello received the 2006 Nobel Prize for discovering the role of RNA interference (RNAi), in the silencing of gene expression.

Starting Materials: The Chemical Elements of Life

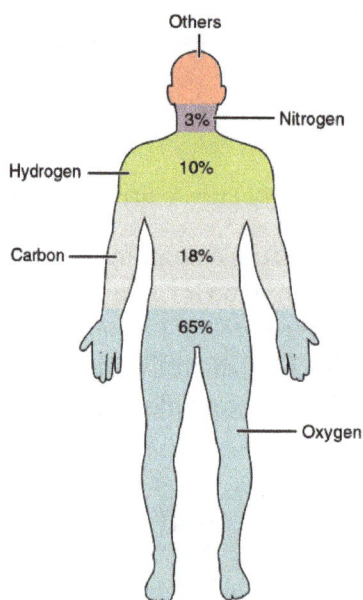

The main elements that compose the human body are shown from most abundant (by mass) to least abundant.

Around two dozen of the 92 naturally occurring chemical elements are essential to various kinds of biological life. Most rare elements on Earth are not needed by life (exceptions being selenium and iodine), while a few common ones (aluminum and titanium) are not used. Most organisms share element needs, but there are a few differences between plants and animals. For example, ocean algae use bromine, but land plants and animals seem to need none. All animals require sodium, but some plants do not. Plants need boron and silicon, but animals may not (or may need ultra-small amounts).

Just six elements—carbon, hydrogen, nitrogen, oxygen, calcium, and phosphorus—make up almost 99% of the mass of living cells, including those in the human body. In addition to the six major elements that compose most of the human body, humans require smaller amounts of possibly 18 more.

Biomolecules

The four main classes of molecules in biochemistry (often called biomolecules) are carbohydrates, lipids, proteins, and nucleic acids. Many biological molecules are polymers: in this terminology, monomers are relatively small micromolecules that are linked together to create large macromolecules known as polymers. When monomers are linked together to synthesize a biological polymer, they undergo a process called dehydration synthesis. Different macromolecules can assemble in larger complexes, often needed for biological activity.

Carbohydrates

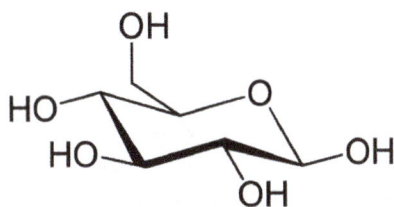

Glucose, a monosaccharide

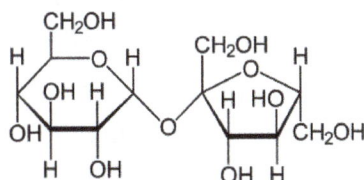

A molecule of sucrose (glucose + fructose), a disaccharide

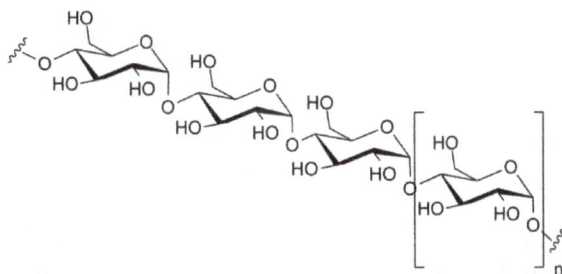

Amylose, a polysaccharide made up of several thousand glucose units

The function of carbohydrates includes energy storage and providing structure. Sugars are carbohydrates, but not all carbohydrates are sugars. There are more carbohydrates on Earth than any other known type of biomolecule; they are used to store energy and genetic information, as well as play important roles in cell to cell interactions and communications.

The simplest type of carbohydrate is a monosaccharide, which among other properties contains carbon, hydrogen, and oxygen, mostly in a ratio of 1:2:1 (generalized formula $C_nH_{2n}O_n$, where n is at least

3). Glucose ($C_6H_{12}O_6$) is one of the most important carbohydrates, others include fructose ($C_6H_{12}O_6$), the sugar commonly associated with the sweet taste of fruits,[a] and deoxyribose ($C_5H_{10}O_4$).

A monosaccharide can switch from the acyclic (open-chain) form to a cyclic form, through a nucleophilic addition reaction between the carbonyl group and one of the hydroxyls of the same molecule. The reaction creates a ring of carbon atoms closed by one bridging oxygen atom. The resulting molecule has an hemiacetal or hemiketal group, depending on whether the linear form was an aldose or a ketose. The reaction is easily reversed, yielding the original open-chain form.

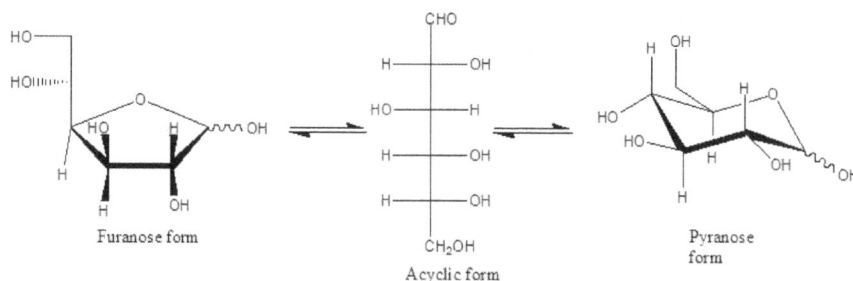

Conversion between the furanose, acyclic, and pyranose forms of D-glucose.

In these cyclic forms, the ring usually has 5 or 6 atoms. These forms are called furanoses and pyranoses, respectively — by analogy with furan and pyran, the simplest compounds with the same carbon-oxygen ring (although they lack the double bonds of these two molecules). For example, the aldohexose glucose may form a hemiacetal linkage between the hydroxyl on carbon 1 and the oxygen on carbon 4, yielding a molecule with a 5-membered ring, called glucofuranose. The same reaction can take place between carbons 1 and 5 to form a molecule with a 6-membered ring, called glucopyranose. Cyclic forms with a 7-atom ring (the same of oxepane), rarely encountered, are called heptoses.

When two monosaccharides undergo dehydration synthesis whereby a molecule of water is released, as two hydrogen atoms and one oxygen atom are lost from the two monosaccharides. The new molecule, consisting of two monosaccharides, is called a *disaccharide* and is conjoined together by a glycosidic or ether bond. The reverse reaction can also occur, using a molecule of water to split up a disaccharide and break the glycosidic bond; this is termed *hydrolysis*. The most well-known disaccharide is sucrose, ordinary sugar (in scientific contexts, called *table sugar* or *cane sugar* to differentiate it from other sugars). Sucrose consists of a glucose molecule and a fructose molecule joined together. Another important disaccharide is lactose, consisting of a glucose molecule and a galactose molecule. As most humans age, the production of lactase, the enzyme that hydrolyzes lactose back into glucose and galactose, typically decreases. This results in lactase deficiency, also called *lactose intolerance*.

When a few (around three to six) monosaccharides are joined, it is called an *oligosaccharide* (*oligo-* meaning "few"). These molecules tend to be used as markers and signals, as well as having some other uses. Many monosaccharides joined together make a polysaccharide. They can be joined together in one long linear chain, or they may be branched. Two of the most common polysaccharides are cellulose and glycogen, both consisting of repeating glucose monomers. Examples are *Cellulose* which is an important structural component of plant's cell walls, and *glycogen*, used as a form of energy storage in animals.

Sugar can be characterized by having reducing or non-reducing ends. A reducing end of a carbohydrate is a carbon atom that can be in equilibrium with the open-chain aldehyde (aldose) or keto form (ketose). If the joining of monomers takes place at such a carbon atom, the free hydroxy group of the pyranose or furanose form is exchanged with an OH-side-chain of another sugar, yielding a full acetal. This prevents opening of the chain to the aldehyde or keto form and renders the modified residue non-reducing. Lactose contains a reducing end at its glucose moiety, whereas the galactose moiety form a full acetal with the C4-OH group of glucose. Saccharose does not have a reducing end because of full acetal formation between the aldehyde carbon of glucose (C1) and the keto carbon of fructose (C2).

Lipids

Lipids comprises a diverse range of molecules and to some extent is a catchall for relatively water-insoluble or nonpolar compounds of biological origin, including waxes, fatty acids, fatty-acid derived phospholipids, sphingolipids, glycolipids, and terpenoids (e.g., retinoids and steroids). Some lipids are linear aliphatic molecules, while others have ring structures. Some are aromatic, while others are not. Some are flexible, while others are rigid.

Lipids are usually made from one molecule of glycerol combined with other molecules. In triglycerides, the main group of bulk lipids, there is one molecule of glycerol and three fatty acids. Fatty acids are considered the monomer in that case, and may be saturated (no double bonds in the carbon chain) or unsaturated (one or more double bonds in the carbon chain).

Most lipids have some polar character in addition to being largely nonpolar. In general, the bulk of their structure is nonpolar or hydrophobic ("water-fearing"), meaning that it does not interact well with polar solvents like water. Another part of their structure is polar or hydrophilic ("water-loving") and will tend to associate with polar solvents like water. This makes them amphiphilic molecules (having both hydrophobic and hydrophilic portions). In the case of cholesterol, the polar group is a mere -OH (hydroxyl or alcohol). In the case of phospholipids, the polar groups are considerably larger and more polar, as described below.

Lipids are an integral part of our daily diet. Most oils and milk products that we use for cooking and eating like butter, cheese, ghee etc., are composed of fats. Vegetable oils are rich in various polyunsaturated fatty acids (PUFA). Lipid-containing foods undergo digestion within the body and are broken into fatty acids and glycerol, which are the final degradation products of fats and lipids. Lipids, especially phospholipids, are also used in various pharmaceutical products, either as co-solubilisers (e.g., in parenteral infusions) or else as drug carrier components (e.g., in a liposome or transfersome).

Proteins

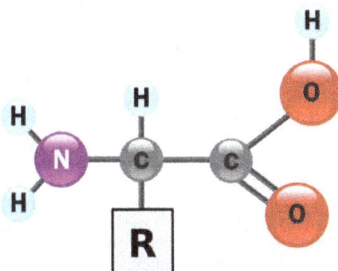

The general structure of an α-amino acid, with the amino group on the left and the carboxyl group on the right.

Proteins are very large molecules – macro-biopolymers – made from monomers called amino acids. An amino acid consists of a carbon atom bound to four groups. One is an amino group, —NH_2, and one is a carboxylic acid group, —COOH (although these exist as —NH_3^+ and —COO^- under physiologic conditions). The third is a simple hydrogen atom. The fourth is commonly denoted "—R" and is different for each amino acid. There are 20 standard amino acids, each containing a carboxyl group, an amino group, and a side-chain (known as an "R" group). The "R" group is what makes each amino acid different, and the properties of the side-chains greatly influence the overall three-dimensional conformation of a protein. Some amino acids have functions by themselves or in a modified form; for instance, glutamate functions as an important neurotransmitter. Amino acids can be joined via a peptide bond. In this dehydration synthesis, a water molecule is removed and the peptide bond connects the nitrogen of one amino acid's amino group to the carbon of the other's carboxylic acid group. The resulting molecule is called a *dipeptide*, and short stretches of amino acids (usually, fewer than thirty) are called *peptides* or polypeptides. Longer stretches merit the title *proteins*. As an example, the important blood serum protein albumin contains 585 amino acid residues.

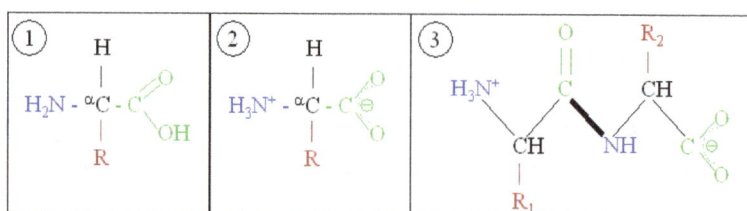

Generic amino acids (1) in neutral form, (2) as they exist physiologically, and (3) joined together as a dipeptide.

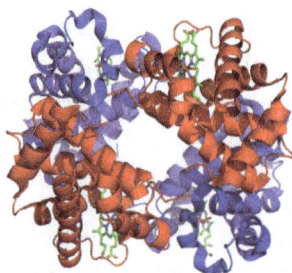

A schematic of hemoglobin. The red and blue ribbons represent the protein globin; the green structures are the heme groups.

Some proteins perform largely structural roles. For instance, movements of the proteins actin and myosin ultimately are responsible for the contraction of skeletal muscle. One property many proteins have is that they specifically bind to a certain molecule or class of molecules—they may be *extremely* selective in what they bind. Antibodies are an example of proteins that attach to one specific type of molecule. In fact, the enzyme-linked immunosorbent assay (ELISA), which uses antibodies, is one of the most sensitive tests modern medicine uses to detect various biomolecules. Probably the most important proteins, however, are the enzymes. Virtually every reaction in a living cell requires an enzyme to lower the activation energy of the reaction. These molecules recognize specific reactant molecules called *substrates*; they then catalyze the reaction between them. By lowering the activation energy, the enzyme speeds up that reaction by a rate of 10^{11} or more; a reaction that would normally take over 3,000 years to complete spontaneously might take less than a second with an enzyme. The enzyme itself is not used up in the process, and is free to catalyze the same reaction with a new set of substrates. Using various modifiers, the activity of the enzyme can be regulated, enabling control of the biochemistry of the cell as a whole.

The structure of proteins is traditionally described in a hierarchy of four levels. The primary structure of a protein simply consists of its linear sequence of amino acids; for instance, "alanine-glycine-tryptophan-serine-glutamate-asparagine-glycine-lysine-...". Secondary structure is concerned with local morphology (morphology being the study of structure). Some combinations of amino acids will tend to curl up in a coil called an α-helix or into a sheet called a β-sheet; some α-helixes can be seen in the hemoglobin schematic above. Tertiary structure is the entire three-dimensional shape of the protein. This shape is determined by the sequence of amino acids. In fact, a single change can change the entire structure. The alpha chain of hemoglobin contains 146 amino acid residues; substitution of the glutamate residue at position 6 with a valine residue changes the behavior of hemoglobin so much that it results in sickle-cell disease. Finally, quaternary structure is concerned with the structure of a protein with multiple peptide subunits, like hemoglobin with its four subunits. Not all proteins have more than one subunit.

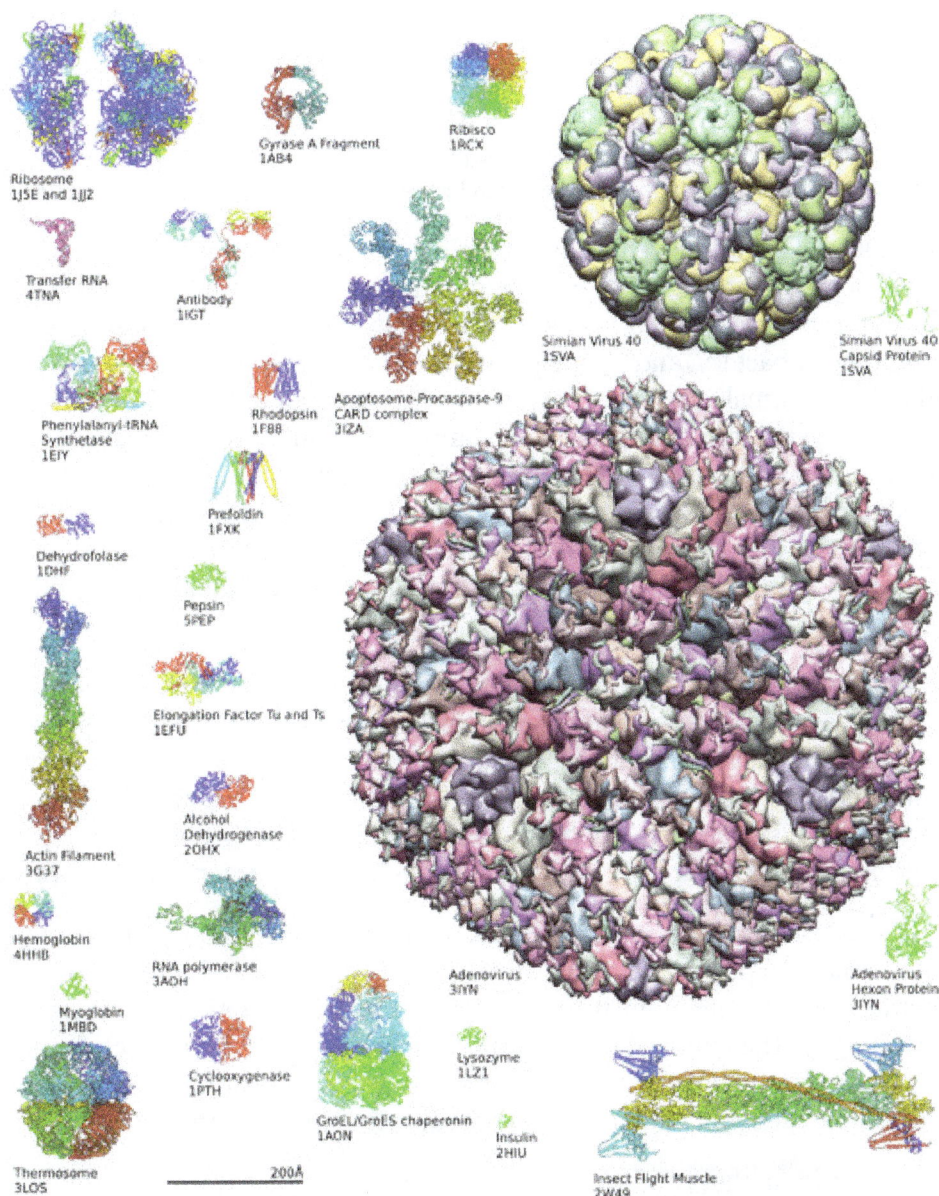

Examples of protein structures from the Protein Data Bank

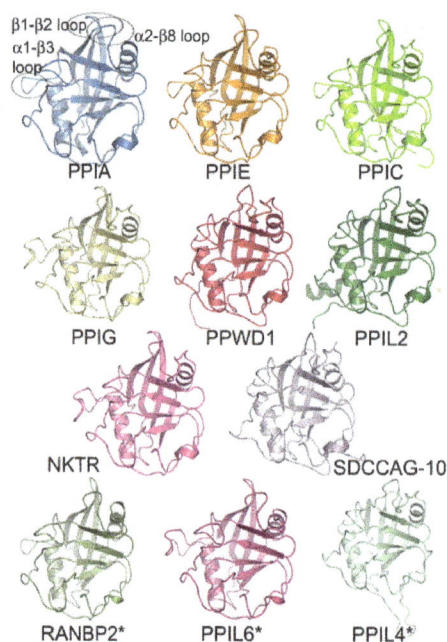

Members of a protein family, as represented by the structures of the isomerase domains.

Ingested proteins are usually broken up into single amino acids or dipeptides in the small intestine, and then absorbed. They can then be joined to make new proteins. Intermediate products of glycolysis, the citric acid cycle, and the pentose phosphate pathway can be used to make all twenty amino acids, and most bacteria and plants possess all the necessary enzymes to synthesize them. Humans and other mammals, however, can synthesize only half of them. They cannot synthesize isoleucine, leucine, lysine, methionine, phenylalanine, threonine, tryptophan, and valine. These are the essential amino acids, since it is essential to ingest them. Mammals do possess the enzymes to synthesize alanine, asparagine, aspartate, cysteine, glutamate, glutamine, glycine, proline, serine, and tyrosine, the nonessential amino acids. While they can synthesize arginine and histidine, they cannot produce it in sufficient amounts for young, growing animals, and so these are often considered essential amino acids.

If the amino group is removed from an amino acid, it leaves behind a carbon skeleton called an α-keto acid. Enzymes called transaminases can easily transfer the amino group from one amino acid (making it an α-keto acid) to another α-keto acid (making it an amino acid). This is important in the biosynthesis of amino acids, as for many of the pathways, intermediates from other biochemical pathways are converted to the α-keto acid skeleton, and then an amino group is added, often via transamination. The amino acids may then be linked together to make a protein.

A similar process is used to break down proteins. It is first hydrolyzed into its component amino acids. Free ammonia (NH_3), existing as the ammonium ion (NH_4^+) in blood, is toxic to life forms. A suitable method for excreting it must therefore exist. Different tactics have evolved in different animals, depending on the animals' needs. Unicellular organisms, of course, simply release the ammonia into the environment. Likewise, bony fish can release the ammonia into the water where it is quickly diluted. In general, mammals convert the ammonia into urea, via the urea cycle.

In order to determine whether two proteins are related, or in other words to decide whether they are homologous or not, scientists use sequence-comparison methods. Methods like sequence alignments and structural alignments are powerful tools that help scientists identify homologies between related molecules. The relevance of finding homologies among proteins goes beyond forming an evolutionary pattern of protein families. By finding how similar two protein sequences are, we acquire knowledge about their structure and therefore their function.

Nucleic Acids

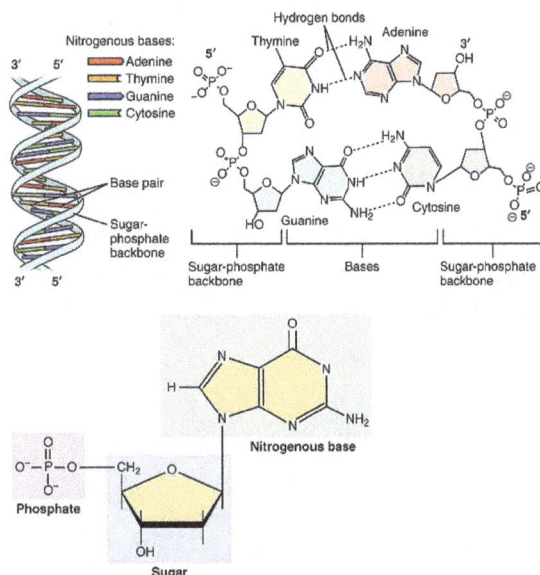

The structure of deoxyribonucleic acid (DNA), the picture shows the monomers being put together.

Nucleic acids, so called because of its prevalence in cellular nuclei, is the generic name of the family of biopolymers. They are complex, high-molecular-weight biochemical macromolecules that can convey genetic information in all living cells and viruses. The monomers are called nucleotides, and each consists of three components: a nitrogenous heterocyclic base (either a purine or a pyrimidine), a pentose sugar, and a phosphate group.

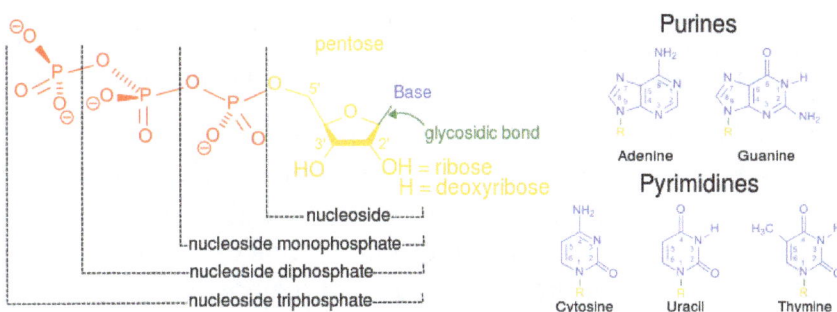

Structural elements of common nucleic acid constituents. Because they contain at least one phosphate group, the compounds marked *nucleoside monophosphate*, *nucleoside diphosphate* and *nucleoside triphosphate* are all nucleotides (not simply phosphate-lacking nucleosides).

The most common nucleic acids are deoxyribonucleic acid (DNA) and ribonucleic acid (RNA). The phosphate group and the sugar of each nucleotide bond with each other to form the backbone of the nucleic acid, while the sequence of nitrogenous bases stores the information. The most common nitrogenous bases are adenine, cytosine, guanine, thymine, and uracil. The nitrogenous bases of each strand of a nucleic acid will form hydrogen bonds with certain other nitrogenous bases in a complementary strand of nucleic acid (similar to a zipper). Adenine binds with thymine and uracil; Thymine binds only with adenine; and cytosine and guanine can bind only with one another.

Aside from the genetic material of the cell, nucleic acids often play a role as second messengers, as well as forming the base molecule for adenosine triphosphate (ATP), the primary energy-carrier molecule found in all living organisms. Also, the nitrogenous bases possible in the two nucleic acids are different: adenine, cytosine, and guanine occur in both RNA and DNA, while thymine occurs only in DNA and uracil occurs in RNA.

Metabolism

Carbohydrates as Energy Source

Glucose is the major energy source in most life forms. For instance, polysaccharides are broken down into their monomers (glycogen phosphorylase removes glucose residues from glycogen). Disaccharides like lactose or sucrose are cleaved into their two component monosaccharides.

Glycolysis (Anaerobic)

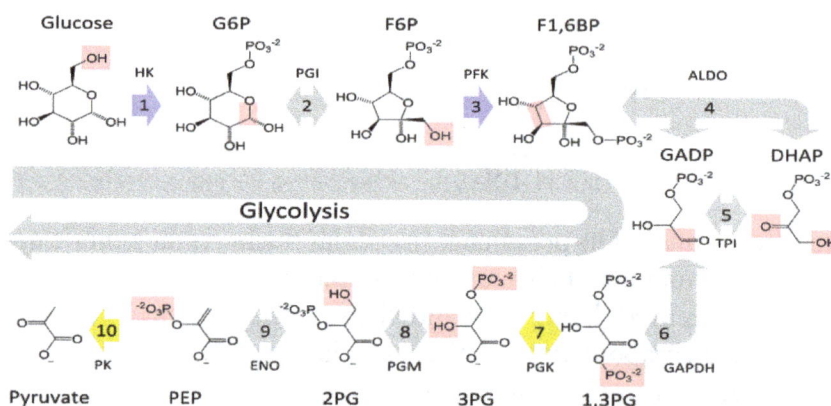

The metabolic pathway of glycolysis converts glucose to pyruvate by via a series of intermediate metabolites. Each chemical modification (red box) is performed by a different enzyme. Steps 1 and 3 consume ATP (blue) and steps 7 and 10 produce ATP (yellow). Since steps 6-10 occur twice per glucose molecule, this leads to a net production of ATP.

Glucose is mainly metabolized by a very important ten-step pathway called glycolysis, the net result of which is to break down one molecule of glucose into two molecules of pyruvate. This also produces a net two molecules of ATP, the energy currency of cells, along with two reducing equivalents of converting NAD^+ (nicotinamide adenine dinucleotide:oxidised form) to NADH (nicotinamide adenine dinucleotide:reduced form). This does not require oxygen; if no oxygen is available (or the cell cannot use oxygen), the NAD is restored by converting the pyruvate to lactate (lactic acid) (e.g., in humans) or to ethanol plus carbon dioxide (e.g., in yeast). Other monosaccharides like galactose and fructose can be converted into intermediates of the glycolytic pathway.

Aerobic

In aerobic cells with sufficient oxygen, as in most human cells, the pyruvate is further metabolized. It is irreversibly converted to acetyl-CoA, giving off one carbon atom as the waste product carbon dioxide, generating another reducing equivalent as NADH. The two molecules acetyl-CoA (from one molecule of glucose) then enter the citric acid cycle, producing two more molecules of ATP, six more NADH molecules and two reduced (ubi)quinones (via $FADH_2$ as enzyme-bound cofactor), and releasing the remaining carbon atoms as carbon dioxide. The produced NADH and quinol molecules then feed into the enzyme complexes of the respiratory chain, an electron transport system transferring the electrons ultimately to oxygen and conserving the released energy in the form of a proton gradient over a membrane (inner mitochondrial membrane in eukaryotes). Thus, oxygen is reduced to water and the original electron acceptors NAD^+ and quinone are regenerated. This is why humans breathe in oxygen and breathe out carbon dioxide. The energy released from transferring the electrons from high-energy states in NADH and quinol is conserved first as proton gradient and converted to ATP via ATP synthase. This generates an additional *28* molecules of ATP (24 from the 8 NADH + 4 from the 2 quinols), totaling to 32 molecules of ATP conserved per degraded glucose (two from glycolysis + two from the citrate cycle). It is clear that using oxygen to completely oxidize glucose provides an organism with far more energy than any oxygen-independent metabolic feature, and this is thought to be the reason why complex life appeared only after Earth's atmosphere accumulated large amounts of oxygen.

Gluconeogenesis

In vertebrates, vigorously contracting skeletal muscles (during weightlifting or sprinting, for example) do not receive enough oxygen to meet the energy demand, and so they shift to anaerobic metabolism, converting glucose to lactate. The liver regenerates the glucose, using a process called gluconeogenesis. This process is not quite the opposite of glycolysis, and actually requires three times the amount of energy gained from glycolysis (six molecules of ATP are used, compared to the two gained in glycolysis). Analogous to the above reactions, the glucose produced can then undergo glycolysis in tissues that need energy, be stored as glycogen (or starch in plants), or be converted to other monosaccharides or joined into di- or oligosaccharides. The combined pathways of glycolysis during exercise, lactate's crossing via the bloodstream to the liver, subsequent gluconeogenesis and release of glucose into the bloodstream is called the Cori cycle.

Relationship to Other "Molecular-scale" Biological Sciences

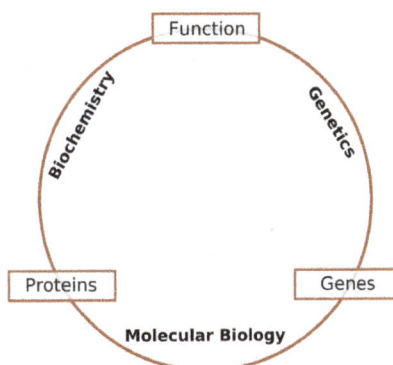

Schematic relationship between biochemistry, genetics, and molecular biology.

Researchers in biochemistry use specific techniques native to biochemistry, but increasingly combine these with techniques and ideas developed in the fields of genetics, molecular biology and biophysics. There has never been a hard-line among these disciplines in terms of content and technique. Today, the terms *molecular biology* and *biochemistry* are nearly interchangeable. The following figure is a schematic that depicts one possible view of the relationship between the fields:

- *Biochemistry* is the study of the chemical substances and vital processes occurring in living organisms. Biochemists focus heavily on the role, function, and structure of biomolecules. The study of the chemistry behind biological processes and the synthesis of biologically active molecules are examples of biochemistry.

- *Genetics* is the study of the effect of genetic differences on organisms. Often this can be inferred by the absence of a normal component (e.g., one gene). The study of "mutants" – organisms with a changed gene that leads to the organism being different with respect to the so-called "wild type" or normal phenotype. Genetic interactions (epistasis) can often confound simple interpretations of such "knock-out" or "knock-in" studies.

- *Molecular biology* is the study of molecular underpinnings of the process of replication, transcription and translation of the genetic material. The central dogma of molecular biology where genetic material is transcribed into RNA and then translated into protein, despite being an oversimplified picture of molecular biology, still provides a good starting point for understanding the field. This picture, however, is undergoing revision in light of emerging novel roles for RNA.

- *Chemical biology* seeks to develop new tools based on small molecules that allow minimal perturbation of biological systems while providing detailed information about their function. Further, chemical biology employs biological systems to create non-natural hybrids between biomolecules and synthetic devices (for example emptied viral capsids that can deliver gene therapy or drug molecules).

Organic Reaction

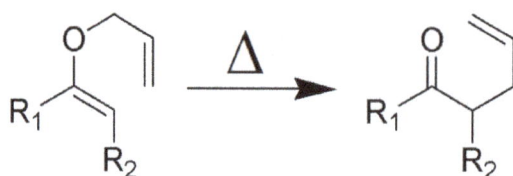

Organic reactions are chemical reactions involving organic compounds. The basic organic chemistry reaction types are addition reactions, elimination reactions, substitution reactions, pericyclic reactions, rearrangement reactions, photochemical reactions and redox reactions. In organic synthesis, organic reactions are used in the construction of new organic molecules. The production of many man-made chemicals such as drugs, plastics, food additives, fabrics depend on organic reactions.

The oldest organic reactions are combustion of organic fuels and saponification of fats to make soap. Modern organic chemistry starts with the Wöhler synthesis in 1828. In the history of the

Nobel Prize in Chemistry awards have been given for the invention of specific organic reactions such as the Grignard reaction in 1912, the Diels-Alder reaction in 1950, the Wittig reaction in 1979 and olefin metathesis in 2005.

Classifications

Organic chemistry has a strong tradition of naming a specific reaction to its inventor or inventors and a long list of so-called named reactions exists, conservatively estimated at 1000. A very old named reaction is the Claisen rearrangement (1912) and a recent named reaction is the Bingel reaction (1993). When the named reaction is difficult to pronounce or very long as in the Corey-House-Posner-Whitesides reaction it helps to use the abbreviation as in the CBS reduction. The number of reactions hinting at the actual process taking place is much smaller, for example the ene reaction or aldol reaction.

Another approach to organic reactions is by type of organic reagent, many of them inorganic, required in a specific transformation. The major types are oxidizing agents such as osmium tetroxide, reducing agents such as Lithium aluminium hydride, bases such as lithium diisopropylamide and acids such as sulfuric acid.

Fundamentals

Factors governing organic reactions are essentially the same as that of any chemical reaction. Factors specific to organic reactions are those that determine the stability of reactants and products such as conjugation, hyperconjugation and aromaticity and the presence and stability of reactive intermediates such as free radicals, carbocations and carbanions.

An organic compound may consist of many isomers. Selectivity in terms of regioselectivity, diastereoselectivity and enantioselectivity is therefore an important criterion for many organic reactions. The stereochemistry of pericyclic reactions is governed by the Woodward–Hoffmann rules and that of many elimination reactions by the Zaitsev's rule.

Organic reactions are important in the production of pharmaceuticals. In a 2006 review it was estimated that 20% of chemical conversions involved alkylations on nitrogen and oxygen atoms, another 20% involved placement and removal of protective groups, 11% involved formation of new carbon-carbon bond and 10% involved functional group interconversions.

Organic Reactions by Mechanism

There is no limit to the number of possible organic reactions and mechanisms. However, certain general patterns are observed that can be used to describe many common or useful reactions. Each reaction has a stepwise reaction mechanism that explains how it happens, although this detailed description of steps is not always clear from a list of reactants alone. Organic reactions can be organized into several basic types. Some reactions fit into more than one category. For example, some substitution reactions follow an addition-elimination pathway. This overview isn't intended to include every single organic reaction. Rather, it is intended to cover the basic reactions.

Reaction type	Subtype	Comment
Addition reactions	electrophilic addition	include such reactions as halogenation, hydrohalogenation and hydration.
	nucleophilic addition	
	radical addition	
Elimination reaction		include processes such as dehydration and are found to follow an E1, E2 or E1cB reaction mechanism
Substitution reactions	nucleophilic aliphatic substitution	with S_N1, S_N2 and $S_N i$ reaction mechanisms
	nucleophilic aromatic substitution	
	nucleophilic acyl substitution	
	electrophilic substitution	
	electrophilic aromatic substitution	
	radical substitution	
Organic redox reactions		are redox reactions specific to organic compounds and are very common.
Rearrangement reactions	1,2-rearrangements	
	pericyclic reactions	
	metathesis	

In condensation reactions a small molecule, usually water, is split off when two reactants combine in a chemical reaction. The opposite reaction, when water is consumed in a reaction, is called hydrolysis. Many Polymerization reactions are derived from organic reactions. They are divided into addition polymerizations and step-growth polymerizations.

In general the stepwise progression of reaction mechanisms can be represented using arrow pushing techniques in which curved arrows are used to track the movement of electrons as starting materials transition to intermediates and products.

Organic Reactions by Functional Groups

Organic reactions can be categorized based on the type of functional group involved in the reaction as a reactant and the functional group that is formed as a result of this reaction. For example, in the Fries rearrangement the reactant is an ester and the reaction product an alcohol.

Other Organic Reaction Classification

In heterocyclic chemistry, organic reactions are classified by the type of heterocycle formed with respect to ring-size and type of heteroatom. Reactions are also categorized by the change in the carbon framework. Examples are ring expansion and ring contraction, homologation reactions, polymerization reactions, insertion reactions, ring-opening reactions and ring-closing reactions.

Organic reactions can also be classified by the type of bond to carbon with respect to the element involved. More reactions are found in organosilicon chemistry, organosulfur chemistry, organophosphorus chemistry and organofluorine chemistry. With the introduction of carbon-metal bonds the field crosses over to organometallic chemistry.

CH																	He
CLi	CBe											CB	CC	CN	CO	CF	Ne
CNa	CMg											CAl	CSi	CP	CS	CCl	CAr
CK	CCa	CSc	CTi	CV	CCr	CMn	CFe	CCo	CNi	CCu	CZn	CGa	CGe	CAs	CSe	CBr	CKr
CRb	CSr	CY	CZr	CNb	CMo	CTc	CRu	CRh	CPd	CAg	CCd	CIn	CSn	CSb	CTe	CI	CXe
CCs	CBa		CHf	CTa	CW	CRe	COs	CIr	CPt	CAu	CHg	CTl	CPb	CBi	CPo	CAt	Rn
Fr	CRa		Rf	Db	CSg	Bh	Hs	Mt	Ds	Rg	Cn	Nh	Fl	Mc	Lv	Ts	Og

↓

CLa	CCe	CPr	CNd	CPm	CSm	CEu	CGd	CTb	CDy	CHo	CEr	CTm	CYb	CLu	
Ac	CTh	CPa	CU	CNp	CPu	CAm	CCm	CBk	CCf	CEs	Fm	Md	No	Lr	

Chemical bonds to carbon	
Core organic chemistry	Many uses in chemistry
Academic research, but no widespread use	Bond unknown

Biochemical and Organic Reaction

The nature's synthesis of complex biological molecules, like proteins, DNA with the help of several enzymes and coenzymes can be correlated with organic synthesis in laboratory. However, Nature's enzymatic synthesis is differed from simple organic chemistry by its degree of complexities and a higher degree of stereoregularity and stereospecificity. It is for that reason why some biological transformations are not easily carried out in test tube in chemistry laboratory. Coenzyme biochemistry often leads to unconventional organic chemistry. In this respect, coenzymes are nature's special reagents. Their well defined chemical structures make them ideal molecules to use for developing the concept of structure-function relationships by bioorganic chemistry approaches as is discussed earlier under biomimetic chemistry section.

Catabolic Processes-Analogy with Organic Chemistry

As an elaborative example, we discussed here the Metabolism reaction in biology and the analogous reactions in Organic Chemistry. In catabolic processes break down and oxidation of larger molecules to produce smaller molecules and energy is take place. The first, beta-oxidation, is a key part of the process by which fatty acids are broken down to acetate.

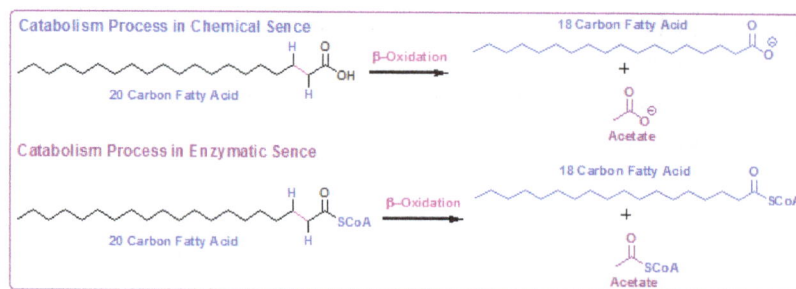

The catabolic process and the analogous organic reaction.

From this we can see that the outcome of a beta-oxidation event is that two carbon atoms are cleaved from a fatty acid. The bond broken is between the alpha and beta carbons. The gamma carbon shows up in the product as a carboxylic acid. This carboxylic acid, two carbons shorter than its parent, can be shortened by another trip through the beta-oxidation process, with the production of another molecule of acetate and a new fatty acid, again two carbons shorter.

Below is the full mechanistic path of the catabolism process. Thus, prior to the commencement of the actual beta-oxidation cycle, the carboxylic acid end of the fatty acid is esterified with the SH group of coenzyme-A.

The first reaction results in the removal of hydrogen atoms from the alpha and beta carbon atoms. Its effects are opposite to those of hydrogenation of a double bond. The removal of hydrogen atoms makes this reaction as an oxidation reaction similar to organic chemistry. The oxidizing agent here, in biology, is FAD (flavine adenine dinucleotide).

Next, water is added to the alkene π-bond which results from the first reaction. This is analogous to the addition of water to alkenes, a popular addition reaction in organic chemistry. Since we know that a carbon alpha to a carbonyl group is a rather nuclophilic place similar to the enolate organic chemistry, it makes sense that the electrophilic hydrogen from water would add there and the nucleophilic -OH would add at the beta carbon.

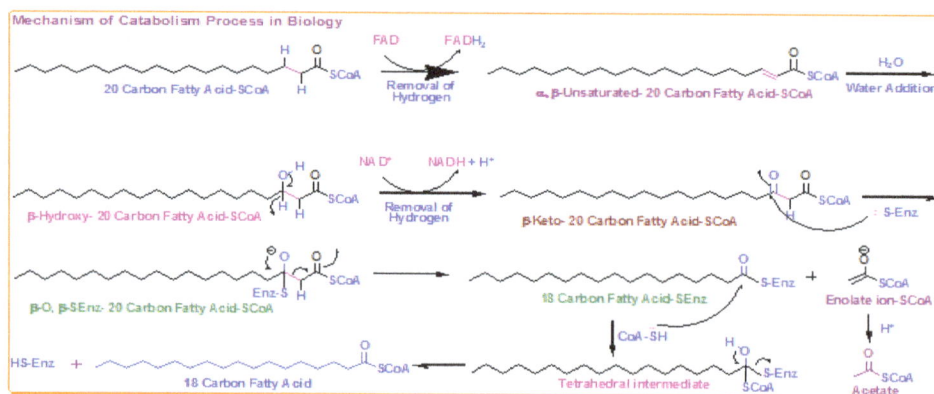

Full mechanism of catabolic process to show analogy with organic reaction.

The third step is nothing but an oxidation of a secondary alcohol to a ketone. This is a reaction for which we used chromium(VI) oxidizing reagents in organic chemistry, on the contrary, in biology, the oxidizing agent is NAD+ (nicotine adenine dinucleotide cation). The hydrogen attached to the OH-bearing carbon is transferred to the NAD+, and the -OH hydrogen comes away as an H+.

In the final step actual cleavage of the C-C bond between the beta-carbon and the gamma carbonyl group takes place. If we think the final step in reverse way, we can see a pattern which is identical to the Claisen condensation reaction in organic chemistry.

Thus we can conclude that the final step in the beta-oxidation cycle is a reverse Claisen condensation. Like the Claisen condensation itself, this step is possible because the enolate ion obtained as a result of breakage of the acetate fragment which is stabilized by resonance. This enolate ion is neutralized by a proton source. The shortened fatty acid is released from the enzyme as CoASH replaces the sulfur of the enzyme. This last step goes through a tetrahedral intermediate similar to organic chemistry as we would expect for a reaction which converts one carboxylic acid derivative to another.

The acetyl-coenzyme-A formed in this cycle enters the tricarboxylic acid cycle where it is oxidized to two molecules of CO_2. The NADH and FADH2 produced in beta-oxidation and the tricarboxylic acid cycle enter a process called oxidative phosphorylation which results in the formation of ATP (adenosine triphosphate) for use in providing energy within the cell.

At last we can conclude that, all the steps catalyzed by enzymes are similar what we can carry out in organic synthesis in laboratory. Overall, that the reactions all occur at the carboxylic acid end of the molecule rather than at the CH_3 end is not surprising, since a fundamental idea of organic chemistry is that reactions occur at functional groups rather than elsewhere.

Glycolysis and its' Analogy with Organic Chemistry

Glycolysis is the conversion of glucose to pyruvate. In the larger scheme of things the pyruvate produced is then converted to acetate, which like the acetate from beta-oxidation of fatty acids, enters the tricarboxylic acid cycle. In this biological process also one can find analogous reaction in organic chemistry. In schematic form the 10 steps of glycolysis are shown along with the analogy with organic chemistry has been elaborated.

Step 1: Glucose is phosphorylated by ATP to form a sugar phosphate. The negative charge of the phosphate prevents passage of the sugar phosphate through the plasma membrane, trapping glucose inside the cell.

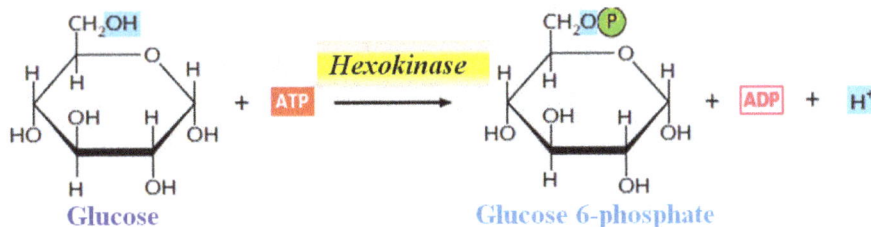

Analogy with Organic Chemistry: Step 1 is an esterification reaction in organic chemistry. It resembles the conversion of an alcohol to an ester by an acid chloride in organic chemistry. Here, ATP is used rather than an acid chloride, and the result is an ester of phosphoric acid rather than an ester of a carboxylic acid.

Step 2: A readily reversible rearrangement of the chemical structure (isomerization) moves the carbonyl oxygen from carbon 1 to carbon 2, forming a ketose from an aldose sugar.

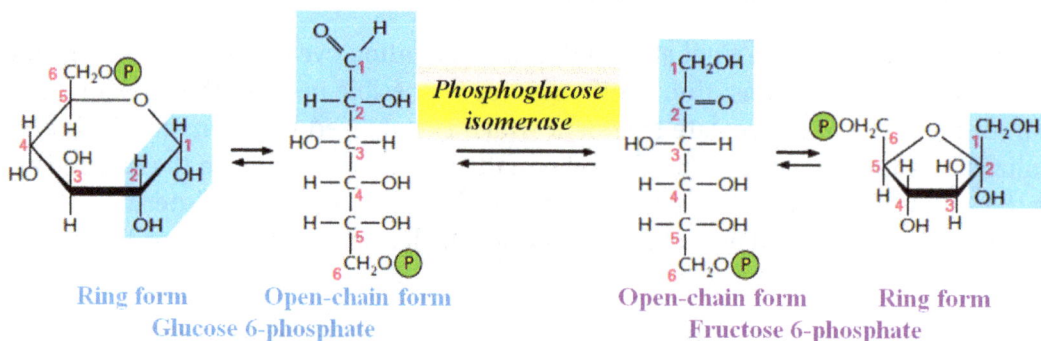

Phosphoglucose isomerase

Ring form — Open-chain form — Open-chain form — Ring form
Glucose 6-phosphate — Fructose 6-phosphate

Analogy with Organic Chemistry: Step 2 is formally an internal oxidation-reduction reaction. The carbonyl group of glucose is reduced to a primary alcohol while the -OH group at C-2 of glucose is oxidized to a ketone. The reaction goes through an enol form similar to organic transformations.

Step 3: The new hydroxyl group on carbon 1 is phosphorylated by ATP, in preparation for the formation of two three-carbon sugar phosphates. The entry of sugars into glycolysis is controlled at this step, through regulation of the enzyme phosphofructokinase.

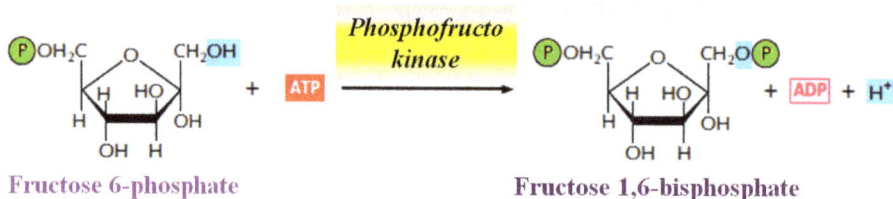

Phosphofructo kinase

Fructose 6-phosphate + ATP → Fructose 1,6-bisphosphate + ADP + H^+

Analogy with Organic Chemistry: Step 3 is similar to Step 1, the formation of a phosphate ester, called phosphorylation.

Step 4: The six carbon sugar is cleaved to produce two three-carbon molecules. Only the glyceraldehyde 3-phosphate can proceed immediately through glycolysis.

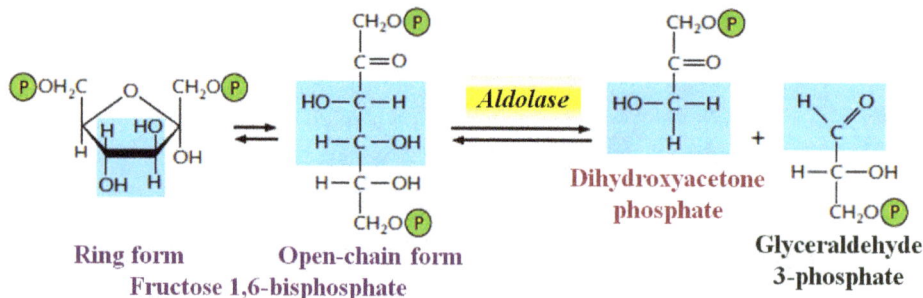

Aldolase

Ring form — Open-chain form — Dihydroxyacetone phosphate + Glyceraldehyde 3-phosphate
Fructose 1,6-bisphosphate

Analogy with Organic Chemistry: Step 4 involves cleavage of a carbon-carbon bond and is the reaction in which the carbon skeleton changes from a six carbon chain to two three carbon chains. Importantly one can notice that the bond, Ca-Cb, adjacent to the carbonyl group is broken. This suggests that enol and enolate reactivity is going to be important. Like we did with the beta-oxidation of fatty acids, we can see a familiar reaction if we think about this step in the reverse direction. If we look at it as a bond-making reaction instead of bond breaking, then it is easier to say that this step is nothing but analogous to a *reverse aldol addition*.

The sequence of mechanistic steps is:

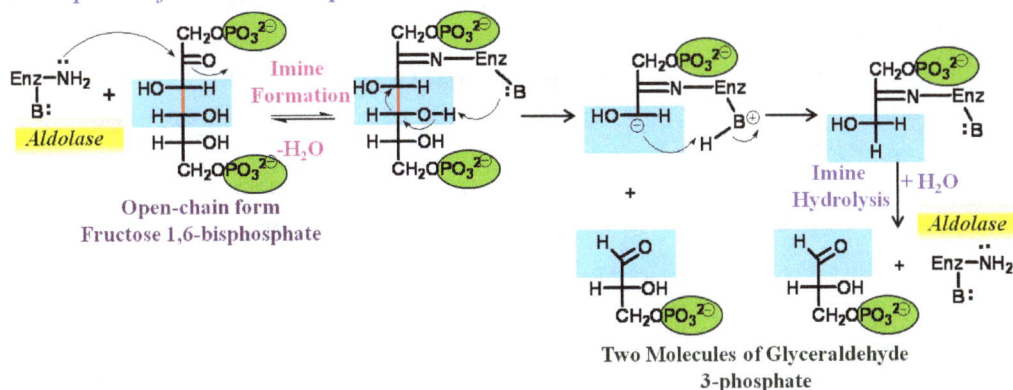

Open-chain form
Fructose 1,6-bisphosphate

Two Molecules of Glyceraldehyde
3-phosphate

Thus, the reverse aldol steps in this reaction are preceded by conversion of the carbonyl group of fructose 1,6-bisphosphate to an imine, and followed by the hydrolysis of that imine back to a carbonyl group and the amino group. The amino group is attached to the enzyme which catalyzes this reaction, and the formation of the imine helps anchoring the fructose 1,6-bisphosphate molecule to the enzyme in the proper position to bring the bases and acids on the enzyme to the right location for the mechanism to go forward.

Step 5: The other product of step 4, dihydroxyacetone phosphate, is isomerized to form glyceraldehyde 3-phosphate.

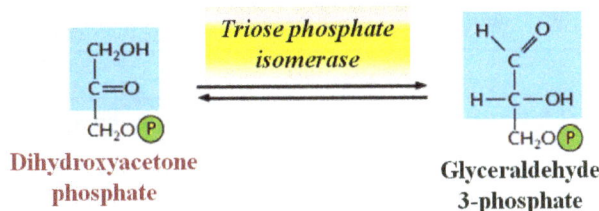

Dihydroxyacetone phosphate

Triose phosphate isomerase

Glyceraldehyde 3-phosphate

Analogy with Organic Chemistry: **Step 5** converts dihydroxy acetone phosphate to glyceraldehyde 3-phosphate. This is like the reverse of **Step 2** in that it goes through an enol intermediate to oxidize an -OH group to a carbonyl on one atom and reduce a carbonyl to an -OH on the adjacent carbon. At this point in glycolysis a glucose molecule has been converted to two molecules of glyceraldehyde 3-phosphate.

Step 6: The two molecules of glyceraldehyde 3-phosphate are oxidized. The energy generation phase of glycolysis begins, as NADH and a new high-energy anhydride linkage to phosphate are formed.

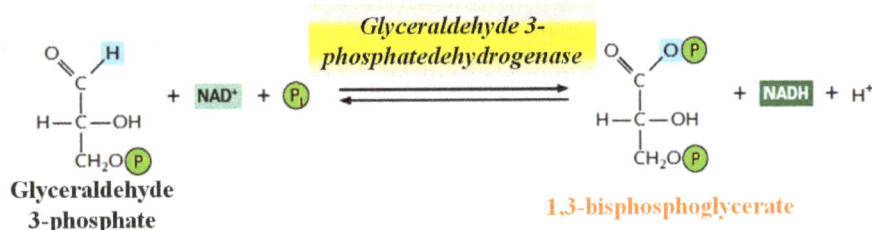

Glyceraldehyde 3-phosphate

Glyceraldehyde 3-phosphate dehydrogenase

$+ NAD^+ + P_i$

$+ NADH + H^+$

1,3-bisphosphoglycerate

Analogy with Organic Chemistry: **Step 6** is an oxidation of the aldehyde in glyceraldehyde 3-phosphate to a carboxylic acid. The oxidizing agent is NAD^+, which is reduced to NADH. The process also includes formation of an anhydride between the newly formed carboxylic acid and a molecule of monohydrogen phosphate ion. This anhydride is quite reactive (since phosphate is a stable anion, it is a good leaving group much like chloride) similar to the reactivity of an acid chloride in organic chemistry.

Step 7: The transfer to ADP of the high energy phosphate group that was generated in **step 6** forms ATP.

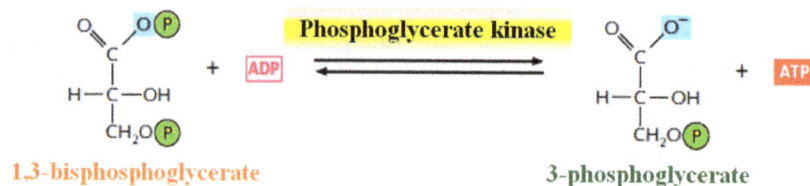

1,3-bisphosphoglycerate   **3-phosphoglycerate**

Analogy with Organic Chemistry: Step 7 exploits the reactivity of this mixed anhydride by using it to transfer the phosphate to ADP. The resulting ATP is used as an energy source for many cell processes, so this step in glycolysis directly produces energy for the cell.

Step 8: The remaining phosphate ester linkage in 3-phosphoglycerate, which has a relatively low free energy of hydrolysis, is moved from carbon 3 to carbon 2 to form 2-phosphoglycerate.

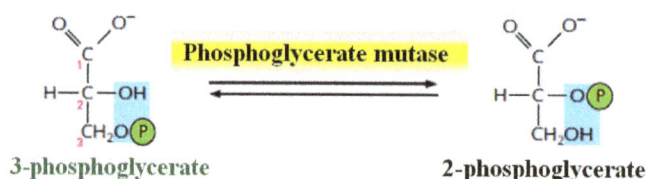

3-phosphoglycerate   **2-phosphoglycerate**

Analogy with Organic Chemistry: Step 8 shifts the phosphate group from the -OH on carbon 3 to the -OH on carbon 2. It is like an ester hydrolysis at one carbon and an esterification at the other.

Step 9: The removal of water from 2-phosphoglycerate creates a high-energy enol phosphate linkage.

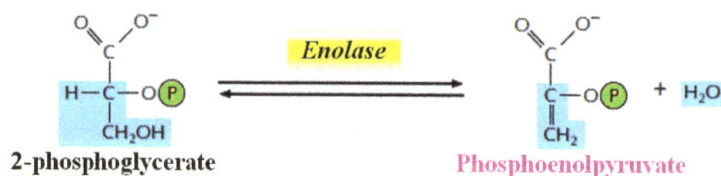

2-phosphoglycerate   **Phosphoenolpyruvate**

Analogy with Organic Chemistry: Step 9 is a dehydration reaction. It resembles the dehydration of cyclohexanol to cyclohexene in organic chemistry, although the conditions in a cell are much milder since enzyme efficiencies make it unnecessary to employ strong acids and heat. The product is an enol phosphate (the phosphoric acid of a ketone).

Step 10: The transfer to ADP of the high-energy phosphate group that was generated in **step 9** forms ATP, thus, completing the glycolysis process.

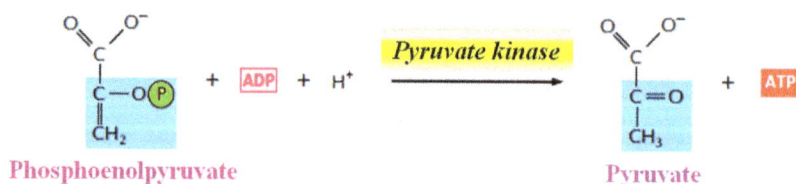

Phosphoenolpyruvate   **Pyruvate**

Analogy with Organic Chemistry: In **Step 10** two things happen. Like **step 7**, transfers of a phosphate to ADP is taking place and thus ATP is formed. Also, the produced enol form of pyruvate rapidly equilibrates to the keto form in this step.

NET RESULT OF GLYCOLYSIS: If we add up all the balanced reactions, we find that one glucose molecule, two ADP molecules, two NAD^+ molecules, and two hydrogen phosphate molecules have been converted to two pyruvates, two ATP's, two NADH's and two hydronium ions.

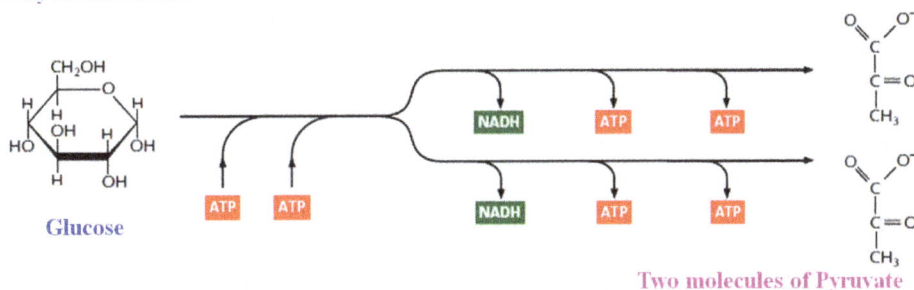

Glucose

Two molecules of Pyruvate

FATE OF PYRUVATE: Fate of pyruvate depends on the local conditions. In muscle tissue there may be a limited supply of oxygen due to exercise or hard work. This means that the NADH produced in glycolysis not oxidized fully back to NAD^+. Glycolysis requires NAD^+. So, it's production and the production of energy in the form of ATP, thus, is also stopped. This can be circumvented by lactate fermentation, which reduces pyrvate to lactate and oxidizes NADH to NAD^+. Energy production can continue until the build-up of lactate and acid. As a result of this reaction muscle exhaustion takes place.

In yeasts, in the absence of oxygen, fermentation to alcohol occurs. This is the basis of the fermentation processes which produce beverage alcohol in beer and wine. The by product is CO_2 which is responsible for the carbonation of beer and sparkling wines. If the yeast is used in baking, the CO_2 expands the dough to produce the rise of bread dough. In this case the alcohol is the byproduct and it is largely driven off by the heat of baking. It is also responsible for much of the pleasant odour of baking bread.

If there is a good oxygen supply, pyruvate is oxidized to acetyl CoA and CO_2. As we have seen now several times, the oxidizing agent is NAD^+ which is reduced to NADH. This reaction is more complex than it looks at first glance.

Pyruvic Acid

Pyruvic acid ($CH_3COCOOH$) is the simplest of the alpha-keto acids, with a carboxylic acid and a ketone functional group. Pyruvate, the conjugate base, CH_3COCOO^-, is a key intermediate in several metabolic pathways.

Pyruvic acid can be made from glucose through glycolysis, converted back to carbohydrates (such as glucose) via gluconeogenesis, or to fatty acids through a reaction with acetyl-CoA. It can also be used to construct the amino acid alanine and can be converted into ethanol or lactic acid via fermentation.

Pyruvic acid supplies energy to cells through the citric acid cycle (also known as the Krebs cycle) when oxygen is present (aerobic respiration), and alternatively ferments to produce lactate when oxygen is lacking (fermentation).

Chemistry

In 1834, Théophile-Jules Pelouze distilled both tartaric acid (L-tartaric acid) and racemic acid (a mix of D- and L-tartaric acid) and isolated pyrotartaric acid (methyl succinic acid) and another acid that Jöns Jacob Berzelius characterized the following year and named pyruvic acid. Pyruvic acid is a colorless liquid with a smell similar to that of acetic acid and is miscible with water. In the laboratory, pyruvic acid may be prepared by heating a mixture of tartaric acid and potassium hydrogen sulfate, by the oxidation of propylene glycol by a strong oxidizer (e.g., potassium permanganate or bleach), or by the hydrolysis of acetyl cyanide, formed by reaction of acetyl chloride with potassium cyanide:

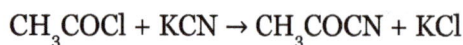

$$CH_3COCl + KCN \rightarrow CH_3COCN + KCl$$

$$CH_3COCN \rightarrow CH_3COCOOH$$

Biochemistry

Pyruvate is an important chemical compound in biochemistry. It is the output of the metabolism of glucose known as glycolysis. One molecule of glucose breaks down into two molecules of pyruvate, which are then used to provide further energy, in one of two ways. Pyruvate is converted into acetyl-coenzyme A, which is the main input for a series of reactions known as the Krebs cycle (also known as the citric acid cycle or tricarboxylic acid cycle). Pyruvate is also converted to oxaloacetate by an anaplerotic reaction, which replenishes Krebs cycle intermediates; also, the oxaloacetate is used for gluconeogenesis. These reactions are named after Hans Adolf Krebs, the biochemist awarded the 1953 Nobel Prize for physiology, jointly with Fritz Lipmann, for research into metabolic processes. The cycle is also known as the citric acid cycle or tricarboxylic acid cycle, because citric acid is one of the intermediate compounds formed during the reactions.

If insufficient oxygen is available, the acid is broken down anaerobically, creating lactate in animals and ethanol in plants and microorganisms (and carp). Pyruvate from glycolysis is converted by fermentation to lactate using the enzyme lactate dehydrogenase and the coenzyme NADH in lactate fermentation, or to acetaldehyde (with the enzyme pyruvate decarboxylase) and then to ethanol in alcoholic fermentation.

Pyruvate is a key intersection in the network of metabolic pathways. Pyruvate can be converted into carbohydrates via gluconeogenesis, to fatty acids or energy through acetyl-CoA, to the amino acid alanine, and to ethanol. Therefore, it unites several key metabolic processes.

Pyruvic Acid Production by Glycolysis

In glycolysis, phosphoenolpyruvate (PEP) is converted to pyruvate by pyruvate kinase. This reaction is strongly exergonic and irreversible; in gluconeogenesis, it takes two enzymes, pyruvate carboxylase and PEP carboxykinase, to catalyze the reverse transformation of pyruvate to PEP.

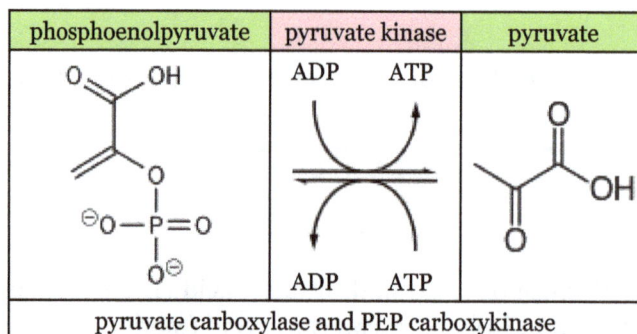

Decarboxylation to Acetyl CoA

Pyruvate decarboxylation by the pyruvate dehydrogenase complex produces acetyl-CoA.

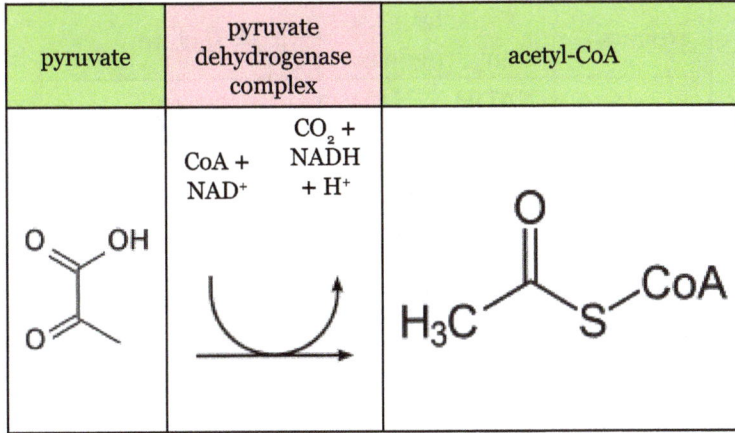

pyruvate	pyruvate dehydrogenase complex	acetyl-CoA
	$CoA +$ NAD^+ $\rightarrow$ $CO_2 +$ $NADH$ $+ H^+$	

Carboxylation to Oxaloacetate

Carboxylation by pyruvate carboxylase produces oxaloacetate.

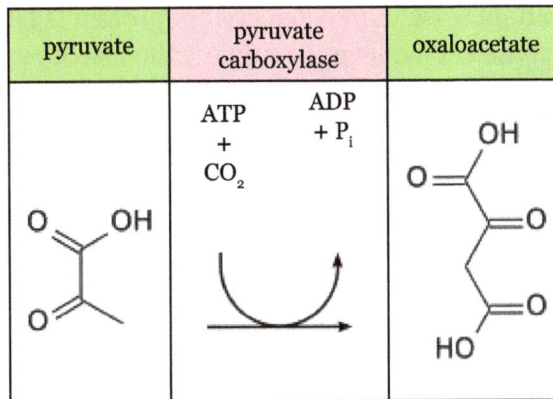

pyruvate	pyruvate carboxylase	oxaloacetate
	ATP $+$ CO_2 $\rightarrow$ ADP $+ P_i$	

Transamination to Alanine

Transamination by alanine transaminase produces alanine.

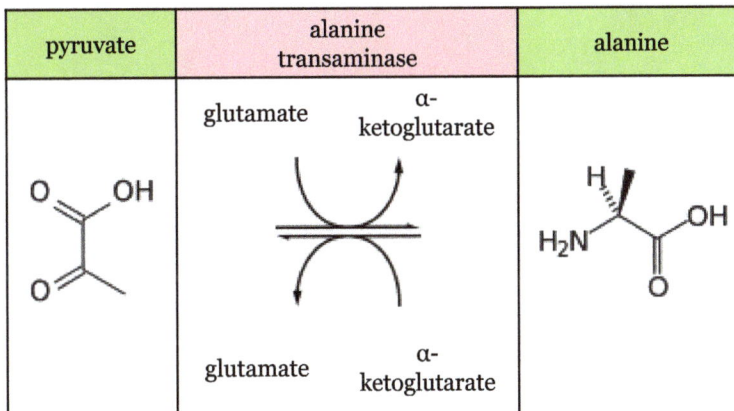

pyruvate	alanine transaminase	alanine
	glutamate → α-ketoglutarate; glutamate → α-ketoglutarate	

Reduction to Lactate

Reduction by lactate dehydrogenase produces lactate.

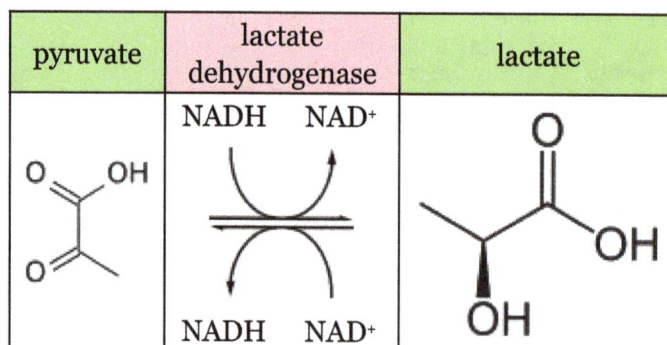

pyruvate	lactate dehydrogenase		lactate

Uses

Pyruvate is sold as a weight-loss supplement, though evidence supporting this use is lacking. A systematic review of six trials found a statistically significant difference in body weight with pyruvate compared to placebo. However, all of the trials had methodological weaknesses and the magnitude of the effect was small. The review also identified adverse events associated with pyruvate such as diarrhea, bloating, gas, and increase in low-density lipoprotein (LDL) cholesterol. The authors concluded that there was insufficient evidence to support the use of pyruvate for weight loss.

There is also *in vitro* as well as *in vivo* evidence in hearts that pyruvate improves metabolism by NADH production stimulation and increases cardiac function.[

5

An Integrated Study of Chemical Biology

Chemical biology, as a subject, focuses on chemistry, physics and biology. It attempts to understand principles related to chemistry in order to apprehend biology and the functions of an organism. The section closely examines the key concepts of chemical biology to provide an extensive understanding of the subject.

Chemical Biology

Chemical biology is a scientific discipline spanning the fields of chemistry, biology, and physics. It involves the application of chemical techniques, tools, and analyses, and often compounds produced through synthetic chemistry, to the study and manipulation of biological systems. Chemical biologists attempt to use chemical principles to modulate systems to either investigate the underlying biology or create new function. Research done by chemical biologists is often closer related to that of cell biology than biochemistry. Biochemists study the chemistry of biomolecules and regulation of biochemical pathways within cells and tissues, e.g. cAMP or cGMP, while chemical biologists deal with novel chemical compounds *applied to* biology.

Introduction

Some forms of chemical biology attempt to answer biological questions by directly probing living systems at the chemical level. In contrast to research using biochemistry, genetics, or molecular biology, where mutagenesis can provide a new version of the organism or cell of interest, chemical biology studies probe systems *in vitro* and *in vivo* with small molecules that have been designed for a specific purpose or identified on the basis of biochemical or cell-based screening.

Chemical biology is one of many interfacial sciences that are characteristic of a general trend away from older, reductionist fields toward those whose goals are to achieve a description of scientific holism. In this sense, it is related to other fields such as proteomics. Chemical biology has scientific, historical and philosophical roots in medicinal chemistry, supramolecular chemistry (particularly host-guest chemistry), bioorganic chemistry, pharmacology, genetics, biochemistry, and metabolic engineering.

Systems of Interest

Proteomics

Proteomics investigates the proteome, the set of expressed proteins at a given time under defined conditions. As a discipline, proteomics has moved past rapid protein identification and has developed into a biological assay for quantitative analysis of complex protein samples by comparing protein changes in differently perturbed systems. Current goals in proteomics include determining protein sequences, abundance and any post-translational modifications. Also of interest are protein–protein interactions, cellular distribution of proteins and understanding protein activity. Another important aspect of proteomics is the advancement of technology to achieve these goals.

Protein levels, modifications, locations, and interactions are complex and dynamic properties. With this complexity in mind, experiments need to be carefully designed to answer specific questions especially in the face of the massive amounts of data that are generated by these analyses. The most valuable information comes from proteins that are expressed differently in a system being studied. These proteins can be compared relative to each other using quantitative proteomics, which allows a protein to be labeled with a mass tag. Proteomic technologies must be sensitive and robust, it is for these reasons, the mass spectrometer has been the workhorse of protein analysis. The high precision of mass spectrometry can distinguish between closely related species and species of interest can be isolated and fragmented within the instrument. Its applications to protein analysis was only possible in the late 1980s with the development of protein and peptide ionization with minimal fragmentation. These breakthroughs were ESI and MALDI. Mass spectrometry technologies are modular and can be chosen or optimized to the system of interest.

Chemical biologists are poised to impact proteomics through the development of techniques, probes and assays with synthetic chemistry for the characterization of protein samples of high complexity. These approaches include the development of enrichment strategies, chemical affinity tags and probes.

Enrichment Techniques

Samples for Proteomics contain a myriad of peptide sequences, the sequence of interest may be highly represented or of low abundance. However, for successful MS analysis the peptide should be enriched within the sample. Reduction of sample complexity is achieved through selective enrichment using affinity chromatography techniques. This involves targeting a peptide with a distinguishing feature like a biotin label or a post translational modification. Interesting methods have been developed that include the use of antibodies, lectins to capture glycoproteins, immobilized metal ions to capture phosphorylated peptides and suicide enzyme substrates to capture specific enzymes. Here, chemical biologists can develop reagents to interact with substrates, specifically and tightly, to profile a targeted functional group on a proteome scale. Development of new enrichment strategies is needed in areas like non-ser/thr/tyr phosphorylation sites and other post translational modifications. Other methods of decomplexing samples relies on upstream chromatographic separations.

Affinity Tags

Chemical synthesis of affinity tags has been crucial to the maturation of quantitative proteomics. iTRAQ, Tandem mass tags (TMT) and Isotope-coded affinity tag (ICAT) are protein mass-tags that consist of a covalently attaching group, a mass (isobaric or isotopic) encoded linker and a handle for isolation. Varying mass-tags bind to different proteins as a sort of footprint such that when analyzing cells of differing perturbations, the levels of each protein can be compared relatively after enrichment by the introduced handle. Other methods include SILAC and heavy isotope labeling. These methods have been adapted to identify complexing proteins by labeling a bait protein, pulling it down and analyzing the proteins it has complexed. Another method creates an internal tag by introducing novel amino acids that are genetically encoded in prokaryotic and eukaryotic organisms. These modifications create a new level of control and can facilitate photocrosslinking to probe protein–protein interactions. In addition, keto, acetylene, azide, thioester, boronate, and dehydroalanine- containing amino acids can be used to selectively introduce tags, and novel chemical functional groups into proteins.

Enzyme Probes

To investigate enzymatic activity as opposed to total protein, activity-based reagents have been developed to label the enzymatically active form of proteins. For example, serine hydrolase- and cysteine protease-inhibitors have been converted to suicide inhibitors. This strategy enhances the ability to selectively analyze low abundance constituents through direct targeting. Structures that mimic these inhibitors could be introduced with modifications that will aid proteomic analysis- like an identification handle or mass tag. Enzyme activity can also be monitored through converted substrate. This strategy relies on using synthetic substrate conjugates that contain moieties that are acted upon by specific enzymes. The product conjugates are then captured by an affinity reagent and analyzed. The measured concentration of product conjugate allow the determination of the enzyme velocity. Identification of enzyme substrates (of which there may be hundreds or thousands, many of which unknown) is a problem of significant difficulty in proteomics and is vital to the understanding of signal transduction pathways in cells; techniques for labelling cellular substrates of enzymes is an area chemical biologists can address. A method that has been developed uses "analog-sensitive" kinases to label substrates using an unnatural ATP analog, facilitating visualization and identification through a unique handle.

Glycobiology

While DNA, RNA and proteins are all encoded at the genetic level, there exists a separate system of trafficked molecules in the cell that are not encoded directly at any direct level: sugars. Thus, glycobiology is an area of dense research for chemical biologists. For instance, live cells can be supplied with synthetic variants of natural sugars in order to probe the function of the sugars in vivo. Carolyn Bertozzi, previously at University of California, Berkeley, has developed a method for site-specifically reacting molecules the surface of cells that have been labeled with synthetic sugars.

Combinatorial Chemistry

Chemical biologists used automated synthesis of many diverse compounds in order to experiment with effects of small molecules on biological processes. More specifically, they observe changes in the behaviors of proteins when small molecules bind to them. Such experiments may supposedly lead to discovery of small molecules with antibiotic or chemotherapeutic properties. These approaches are identical to those employed in the discipline of pharmacology.

Molecular Sensing

Chemical biologists are also interested in developing new small-molecule and biomolecule-based tools to study biological processes, often by molecular imaging techniques. The field of molecular sensing was popularized by Roger Tsien's work developing calcium-sensing fluorescent compounds as well as pioneering the use of GFP, for which he was awarded the 2008 Nobel Prize in

Chemistry. Today, researchers continue to utilize basic chemical principles to develop new compounds for the study of biological metabolites and processes.

siRNA-A Tool in Chemical Biology

siRNA or small interfering RNAs owe their origins to the difficulties the scientific community faced utilizing classical and reverse genetics methods in studying gene expression. Disrupting genes to study their functions is not always optimal; neither is mapping mutations back to their genes easy. The whole process is expensive as well as time-consuming, which is why a lot of effort has been devoted to develop methods to silence gene expression in sequence specific manner using nucleic acids. They have the potential to be powerful tools in the field of chemical biology to study the chemistry of gene expression in therapeutic targets of bacteria and viruses.

A number of different types of nucleic acid molecules have already gained prominence because of their potential as therapeutics. They target mRNAs to silence the genes in a sequence specific manner. Oligodeoxyribonucleic acids, ODNs utilize steric interaction to silence gene expression. They can also form triple helices in conjunction with the DNA duplex. Whereas ribozymes can be chemically designed to target specific genes and cleave them in a sequence specific manner. The most promising of these methods however is utilization of short interfering RNA or siRNA to silence gene expression.

siRNA

siRNA or short interfering RNAs exist in nature as a means for the express purpose of controlling gene expression. It was discovered in petunia as a post-transcriptional gene silencing measure. It is the resultant product when a long double-strand RNA of 20 -25 nucleotides length was processed in the cells by the enzyme DICER. The newly synthesized siRNA assemble into endoribonuclease-containing complexes known as RNA-induced silencing complexes (RISCs), unwinding in the process. The activated RISC then binds to the complementary RNA molecules by base pairing interactions between the siRNA strand and the mRNA, which is then cleaved. This mechanism is known as RNA interference or RNAi.

Designing and Synthesizing siRNAs

It is now possible to order siRNAs designed and synthesized with the express purpose of targeting a particular sequence. The ambion website has a lot of information on the optimal design of siRNAs.

siRNAs can be synthesized chemically, or enzymatically. RNase III or DICER can be used to cleave the long dsRNAs to produce siRNAs. However the most expedient method is the use of plasmids to express them in vivo by delivering them into the target cell using vectors. This method allows the siRNAs to be expressed in the target cell stably, over a period of time and overcomes the drawbacks of the transience of their effect. Numerous strategies have been developed in order to deliver the siRNA into the cell efficiently:

1. Electroporation

2. Local and systemic injection: This method was the first success scientists had in silencing genes using siRNAs. They were successfully delivered into highly vascularized tissue in mice through

using high-pressure tail vein injection. Greater than 90% loss in gene expression was observed in the targets.

3. siRNA producing viruses: This method shows great promise in gene therapy, and research is progressing in order to generate recombinant viruses that can produce siRNA in target cells.

4. Small molecules that enhance transdermal penetration: Research in this field is moving at a fast pace in order to synthesize small organic molecules that, if injected in conjunction with siRNAs, can help them penetrate into the target cells.

Biological uses of the RNAi Approach

The principal purpose of studying siRNA mediated RNA interference is probably to investigate gene function.

It is so much easier to make genetic knock-outs by simply introducing sequence-specific siRNAs into cells; multi-copy genes can be silenced in one fell swoop by this method. Creation of double-knockout mutants is also easier and consumes much less time. Using local injections in specific regions of the model organisms also help in creating spatially separated and restricted knockout. siRNAs are also being successfully used to screen whole genomes in organisms such as C. elegans and Drosophilla melanogaster. Even in mammalian systems such as Danio rerio (zebrafish) that usually prove intractable to all gene silencing methods, even dsRNA injection, siRNA can do the job. It is paving a new way in development of therapeutics by identifying human gene orthologs in other species in a remarkably short period of time.

Numerous high-throughput screening approaches are being developed to screen large libraries of cells rapidly in order to identify drug targets. A brief description of few of the screening techniques:

1. Pooled Format Screening: A reagent library of RNAi has to be introduced to the cells so that a particular cell is in one particular reagent. The primary hits are then identified and their identity elucidate by sequencing techniques.

2. Arrayed Format Screening: Each RNAi reagent is placed in separate wells in a plate and multiple manipulations can be done to identify their targets, which are then detected by fluorescence readouts, imaging techniques and other methods as well. Thus the identity of the target cell can be determined through the identity of the reagent in the database.

3. Multiplexed methods: A combination of various assays can be used for high-throughput screening of candidate drug targets. For example, candidate genes can be identified through informatics based methods and then screened against a library of reagents. Many other such methods are being developed in order to make the job of screening therapeutic targets easier.

siRNA Based Therapeutics

The future in this field rests in the development of siRNA-based drugs.

This could prove to be a powerful tool in gene based therapy. Research is now concentrated on developing strategies to design siRNA therapeutics for clinical use. A brief description of some novel strategies for siRNA drug development is provided here:

1. Direct Mutation Targeting: The siRNAs are designed to perfectly match mutant alleles but contain one or more mismatches with wild-type alleles, leading to specific degradation of the matching, mutant transcripts.

2. Indirect Mutation Targeting: The siRNA approach will not work if the mutant alleles are too similar to wild type. So an indirect approach is taken in which siRNAs are designed against disease linked markers such as SNP variations. The ones that are screened as positive are targeted for degradation.

3. Exon-specific targeting: siRNAs are designed to target expressed regions (exons) of the gene.

4. Targeting exon skipped transcripts: If the problem in the gene lies in aberrant splicing post-transcription, siRNA can be designed to target the unnatural exon-exon interface arising as a result of such alternative splicing.

Employing Biology

Many research programs are also focused on employing natural biomolecules to perform a task or act as support for a new chemical method or material. In this regard, researchers have shown that DNA can serve as a template for synthetic chemistry, self-assembling proteins can serve as a structural scaffold for new materials, and RNA can be evolved *in vitro* to produce new catalytic function.

Protein Misfolding and Aggregation as a Cause of Disease

A common form of aggregation is long, ordered spindles called amyloid fibrils that are implicated in Alzheimer's disease and that have been shown to consist of cross-linked beta sheet regions perpendicular to the backbone of the polypeptide. Another form of aggregation occurs with prion proteins, the glycoproteins found with Creutzfeldt–Jakob disease and bovine spongiform encephalopathy. In both structures, aggregation occurs through hydrophobic interactions and water must be excluded from the binding surface before aggregation can occur. A movie of this process can be seen in "Chemical and Engineering News". The diseases associated with misfolded proteins are life-threatening and extremely debilitating, which makes them an important target for chemical biology research.

Through the transcription and translation process, DNA encodes for specific sequences of amino acids. The resulting polypeptides fold into more complex secondary, tertiary, and quaternary structures to form proteins. Based on both the sequence and the structure, a particular protein is conferred its cellular function. However, sometimes the folding process fails due to mutations in the genetic code and thus the amino acid sequence or due to changes in the cell environment (e.g. pH, temperature, reduction potential, etc.). Misfolding occurs more often in aged individuals or in cells exposed to a high degree of oxidative stress, but a fraction of all proteins misfold at some point even in the healthiest of cells.

Normally when a protein does not fold correctly, molecular chaperones in the cell can encourage refolding back into its active form. When refolding is not an option, the cell can also target the protein for degradation back into its component amino acids via proteolytic, lysosomal, or autophagic mechanisms. However, under certain conditions or with certain mutations, the cells can no longer cope with the misfolded protein(s) and a disease state results. Either the protein has a loss-of-function, such as in cystic fibrosis, in which it loses activity or cannot reach its target, or the protein has a gain-of-function, such as with Alzheimer's disease, in which the protein begins to aggregate causing it to become insoluble and non-functional.

Protein misfolding has previously been studied using both computational approaches as well as *in vivo* biological assays in model organisms such as *Drosophila melanogaster* and *C. elegans*. Computational models use a *de novo* process to calculate possible protein structures based on input parameters such as amino acid sequence, solvent effects, and mutations. This method has the shortcoming that the cell environment has been drastically simplified, which limits the factors that influence folding and stability. On the other hand, biological assays can be quite complicated to perform *in vivo* with high-throughput like efficiency and there always remains the question of how well lower organism systems approximate human systems.

Dobson et al. propose combining these two approaches such that computational models based on the organism studies can begin to predict what factors will lead to protein misfolding. Several experiments have already been performed based on this strategy. In experiments on *Drosophila*, different mutations of beta amyloid peptides were evaluated based on the survival rates of the flies as well as their motile ability. The findings from the study show that the more a protein aggregates, the more detrimental the neurological dysfunction. Further studies using transthyretin, a component of cerebrospinal fluid that binds to beta amyloid peptide deterring aggregation but can itself aggregate especially when mutated, indicate that aggregation prone proteins may not aggregate where they are secreted and rather are deposited in specific organs or tissues based on each mutation. Kelly et al. have shown that the more stable, both kinetically and thermodynamically, a misfolded protein is the more likely the cell is to secrete it from the endoplasmic reticulum rather than targeting the protein for degradation. In addition, the more stress that a cell feels from misfolded proteins the more probable new proteins will misfold. These experiments as well as others having begun to elucidate both the intrinsic and extrinsic causes of misfolding as well as how the cell recognizes if proteins have folded correctly.

As more information is obtained on how the cell copes with misfolded proteins, new therapeutic strategies begin to emerge. An obvious path would be prevention of misfolding. However, if protein misfolding cannot be avoided, perhaps the cell's natural mechanisms for degradation can be bolstered to better deal with the proteins before they begin to aggregate. Before these ideas can be realized, many more experiments need to be done to understand the folding and degradation machinery as well as what factors lead to misfolding. More information about protein misfolding and how it relates to disease can be found in the recently published book by Dobson, Kelly, and Rameriz-Alvarado entitled Protein Misfolding Diseases Current and Emerging Principles and Therapies.

Chemical Synthesis of Peptides

In contrast to the traditional biotechnological practice of obtaining peptides or proteins by isolation from cellular hosts through cellular protein production, advances in chemical techniques for the synthesis and ligation of peptides has allowed for the total synthesis of some peptides and proteins. Chemical synthesis of proteins is a valuable tool in chemical biology as it allows for the introduction of non-natural amino acids as well as residue specific incorporation of "posttranslational modifications" such as phosphorylation, glycosylation, acetylation, and even ubiquitination.

These capabilities are valuable for chemical biologists as non-natural amino acids can be used to probe and alter the functionality of proteins, while post translational modifications are widely known to regulate the structure and activity of proteins. Although strictly biological techniques have been developed to achieve these ends, the chemical synthesis of peptides often has a lower technical and practical barrier to obtaining small amounts of the desired protein. Given the widely recognized importance of proteins as cellular catalysts and recognition elements, the ability to precisely control the composition and connectivity of polypeptides is a valued tool in the chemical biology community and is an area of active research.

While chemists have been making peptides for over 100 years, the ability to efficiently and quickly synthesize short peptides came of age with the development of Bruce Merrifield's solid phase peptide synthesis (SPPS). Prior to the development of SPPS, the concept of step-by-step polymer synthesis on an insoluble support was without chemical precedent. The use of a covalently bound insoluble polymeric support greatly simplified the process of peptide synthesis by reducing purification to a simple "filtration and wash" procedure and facilitated a boom in the field of peptide chemistry. The development and "optimization" of SPPS took peptide synthesis from the hands of the specialized peptide synthesis community and put it into the hands of the broader chemistry, biochemistry, and now chemical biology community. SPPS is still the method of choice for linear synthesis of polypeptides up to 50 residues in length and has been implemented in commercially available automated peptide synthesizers. One inherent shortcoming in any procedure that calls for repeated coupling reactions is the buildup of side products resulting from incomplete couplings and side reactions. This places the upper bound for the synthesis of linear polypeptide lengths at around 50 amino acids, while the "average" protein consists of 250 amino acids. Clearly, there was a need for development of "non-linear" methods to allow synthetic access to the average protein.

Although the shortcomings of linear SPPS were recognized not long after its inception, it took until the early 1990s for effective methodology to be developed to ligate small peptide fragments made by SPPS, into protein sized polypeptide chains (for recent review of peptide ligation strategies, see review by Dawson *et al.*). The oldest and best developed of these methods is termed native chemical ligation. Native chemical ligation was unveiled in a 1994 paper from the laboratory of Stephen B. H. Kent. Native chemical ligation involves the coupling of a C-terminal thioester and an N-terminal cysteine residue, ultimately resulting in formation of a "native" amide bond. Further refinements in native chemical ligation have allowed for kinetically controlled coupling of multiple peptide fragments, allowing access to moderately sized peptides such as an HIV-protease dimer and human lysozyme. Even with the successes and attractive features of native chemical ligation, there are still some drawbacks in the utilization of this technique. Some of these drawbacks include the installation and preservation of a reactive C-terminal thioester, the requirement of an N-terminal cysteine residue (which is the second-least-common amino acid in proteins), and the requirement for a sterically unincumbering C-terminal residue.

Other strategies that have been used for the ligation of peptide fragments using the acyl transfer chemistry first introduced with native chemical ligation include expressed protein ligation, sulfurization/desulfurization techniques, and use of removable thiol auxiliaries.

Expressed protein ligation allows for the biotechnological installation of a C-terminal thioester using intein biochemistry, thereby allowing the appendage of a synthetic N-terminal peptide to the recombinantly produced C-terminal portion. This technique allows for access to much larger proteins, as only the N-terminal portion of the resulting protein has to be chemically synthesized. Both sulfurization/desulfurization techniques and the use of removable thiol auxiliaries involve the installation of a synthetic thiol moiety to carry out the standard native chemical ligation chemistry, followed by removal of the auxiliary/thiol. These techniques help to overcome the requirement of an N-terminal cysteine needed for standard native chemical ligation, although the steric requirements for the C-terminal residue are still limiting.

A final category of peptide ligation strategies include those methods not based on native chemical ligation type chemistry. Methods that fall in this category include the traceless Staudinger ligation, azide-alkyne dipolar cycloadditions, and imine ligations.

Major contributors in this field today include Stephen B. H. Kent, Philip E. Dawson, and Tom W. Muir, as well as many others involved in methodology development and applications of these strategies to biological problems.

Protein Design by Directed Evolution

One of the primary goals of protein engineering is the design of novel peptides or proteins with a desired structure and chemical activity. Because our knowledge of the relationship between primary sequence, structure, and function of proteins is limited, rational design of new proteins with enzymatic activity is extremely challenging. Directed evolution, repeated cycles of genetic diversification followed by a screening or selection process, can be used to mimic Darwinian evolution in the laboratory to design new proteins with a desired activity.

Several methods exist for creating large libraries of sequence variants. Among the most widely used are subjecting DNA to UV radiation or chemical mutagens, error-prone PCR, degenerate codons, or recombination. Once a large library of variants is created, selection or screening techniques are used to find mutants with a desired attribute. Common selection/screening techniques include fluorescence-activated cell sorting (FACS), mRNA display, phage display, or *in vitro* compartmentalization. Once useful variants are found, their DNA sequence is amplified and subjected to further rounds of diversification and selection. Since only proteins with the desired activity are selected, multiple rounds of directed evolution lead to proteins with an accumulation beneficial traits.

There are two general strategies for choosing the starting sequence for a directed evolution experiment: *de novo* design and redesign. In a protein design experiment, an initial sequence is chosen at random and subjected to multiple rounds of directed evolution. For example, this has been employed successfully to create a family of ATP-binding proteins with a new folding pattern not found in nature. Random sequences can also be biased towards specific folds by specifying the characteristics (such as polar vs. nonpolar) but not the specific identity of each amino acid in a sequence. Among other things, this strategy has been used to successfully design four-helix bundle proteins. Because it is often thought that a well-defined structure is required for activity, biasing a designed protein towards adopting a specific folded structure is likely to increase the frequency of desirable variants in constructed libraries.

In a protein redesign experiment, an existing sequence serves as the starting point for directed evolution. In this way, old proteins can be redesigned for increased activity or new functions. Protein redesign has been used for protein simplification, creation of new quaternary structures, and topological redesign of a chorismate mutase. To develop enzymes with new activities, one can take advantage of promiscuous enzymes or enzymes with significant side reactions. In this regard, directed evolution has been used on γ-humulene synthase, an enzyme that creates over 50 different sesquiterpenes, to create enzymes that selectively synthesize individual products. Similarly, completely new functions can be selected for from existing protein scaffolds. In one example of this, an RNA ligase was created from a zinc finger scaffold after 17 rounds of directed evolution. This new enzyme catalyzes a chemical reaction not known to be catalyzed by any natural enzyme.

Computational methods, when combined with experimental approaches, can significantly assist both the design and redesign of new proteins through directed evolution. Computation has been used to design proteins with unnatural folds, such as a right-handed coiled coil. These computational approaches could also be used to redesign proteins to selectively bind specific target molecules. By identifying lead sequences using computational methods, the occurrence of functional proteins in libraries can be dramatically increased before any directed evolution experiments in the laboratory.

Manfred T. Reetz, Frances Arnold, Donald Hilvert, and Jack W. Szostak are significant researchers in this field.

Biocompatible Click Cycloaddition Reactions in Chemical Biology

Recent advances in technology have allowed scientists to view substructures of cells at levels of unprecedented detail. Unfortunately these "aerial" pictures offer little information about the mechanics of the biological system in question. To be fully effective, precise imaging systems require a complementary technique that better elucidates the machinery of a cell. By attaching tracking devices (optical probes) to biomolecules *in vivo*, one can learn far more about cell metabolism, molecular transport, cell-cell interactions and many other processes

Bioorthogonal Reactions

Successful labeling of a molecule of interest requires specific functionalization of that molecule to react chemospecifically with an optical probe. For a labeling experiment to be considered robust, that functionalization must minimally perturb the system. Unfortunately, these requirements can often be extremely hard to meet. Many of the reactions normally available to organic chemists in the laboratory are unavailable in living systems. Water- and redox- sensitive reactions would not proceed, reagents prone to nucleophilic attack would offer no chemospecificity, and any reactions with large kinetic barriers would not find enough energy in the relatively low-heat environment of a living cell. Thus, chemists have recently developed a panel of bioorthogonal chemistry that proceed chemospecifically, despite the milieu of distracting reactive materials *in vivo*.

Design of Bioorthogonal Reagents and Bioorthogonal Chemical Reporters

The coupling of an optical probe to a molecule of interest must occur within a reasonably short time frame; therefore, the kinetics of the coupling reaction should be highly favorable. Click chemistry is well suited to fill this niche, since click reactions are, by definition, rapid, spontaneous, selective, and high-yielding. Unfortunately, the most famous "click reaction," a [3+2] cycloaddition between an azide and an acyclic alkyne, is copper-catalyzed, posing a serious problem for use *in vivo* due to copper's toxicity.

The issue of copper toxicity can be alleviated using copper-chelating ligands, enabling copper-catalyzed labeling of the surface of live cells.

To bypass the necessity for a catalyst, the lab of Dr. Carolyn Bertozzi introduced inherent strain into the alkyne species by using a cyclic alkyne. In particular, cyclooctyne reacts with azido-molecules with distinctive vigor. Further optimization of the reaction led to the use of difluorinated cyclooctynes (DIFOs), which increased yield and reaction rate. Other coupling partners discovered by separate labs to be analogous to cyclooctynes include trans cyclooctene, norbornene, and a cyclobutene-functionalized molecule.

Use in Biological Systems

As mentioned above, the use of bioorthogonal reactions to tag biomolecules requires that one half of the reactive "click" pair is installed in the target molecule, while the other is attached to an optical probe. When the probe is added to a biological system, it will selectively conjugate with the target molecule.

The most common method of installing bioorthogonal reactivity into a target biomolecule is through metabolic labeling. Cells are immersed in a medium where access to nutrients is limited to synthetically modified analogues of standard fuels such as sugars. As a consequence, these altered biomolecules are incorporated into the cells in the same manner as their wild-type brethren. The optical probe is then incorporated into the system to image the fate of the altered biomolecules. Other methods of functionalization include enzymatically inserting azides into proteins, and synthesizing phospholipids conjugated to cyclooctynes.

Future Directions

As these bioorthogonal reactions are further optimized, they will likely be used for increasingly complex interactions involving multiple different classes of biomolecules. More complex interactions have a smaller margin for error, so increased reaction efficiency is paramount to continued success in optically probing cellular machinery. Also, by minimizing side reactions, the experimental design of a minimally perturbed living system is closer to being realized.

Discovery of Biomolecules through Metagenomics

The advances in modern sequencing technologies in the late 1990s allowed scientists to investigate DNA of communities of organisms in their natural environments, so-called "eDNA", without culturing individual species in the lab. This metagenomic approach enabled scientists to study a wide selection of organisms that were previously not characterized due in part to an incompetent growth condition. These sources of eDNA include, but are not limited to, soils, ocean, subsurface, hot springs, hydrothermal vents, polar ice caps, hypersaline habitats, and extreme pH environments. Of the many applications of metagenomics, chemical biologists and microbiologists such as Jo Handelsman, Jon Clardy, and Robert M. Goodman who are pioneers of metagenomics, explored metagenomic approaches toward the discovery of biologically active molecules such as antibiotics.

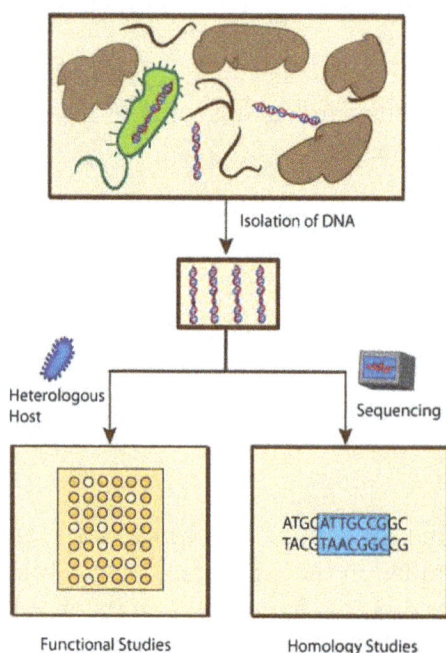

Overview of metagenomic methods

Functional or homology screening strategies have been used to identify genes that produce small bioactive molecules. Functional metagenomic studies are designed to search for specific phenotypes that are associated with molecules with specific characteristics. Homology metagenomic studies, on the other hand, are designed to examine genes to identify conserved sequences that are previously associated with the expression of biologically active molecules.

Functional metagenomic studies enable scientists to discover novel genes that encode biologically active molecules. These assays include top agar overlay assays where antibiotics generate zones of growth inhibition against test microbes, and pH assays that can screen for pH change due to newly synthesized molecules using pH indicator on an agar plate. Substrate-induced gene expression screening (SIGEX), a method to screen for the expression of genes that are induced by chemical compounds, has also been used to search genes with specific functions. These led to the discovery and isolation of several novel proteins and small molecules. For example, the Schipper group identified three eDNA derived AHL lactonases that inhibit biofilm formation of Pseudomonas aeruginosa via functional metagenomic assays. However, these functional screening methods require a good design of probes that detect molecules being synthesized and depend on the ability to express metagenomes in a host organism system.

In contrast, homology metagenomic studies led to a faster discovery of genes that have homologous sequences as the previously known genes that are responsible for the biosynthesis of biologically active molecules. As soon as the genes are sequenced, scientists can compare thousands of bacterial genomes simultaneously. The advantage over functional metagenomic assays is that homology metagenomic studies do not require a host organism system to express the metagenomes, thus this method can potentially save the time spent on analyzing nonfunctional genomes. These also led to the discovery of several novel proteins and small molecules. For example, Banik et al. screened for clones containing genes associated with the synthesis of teicoplanin and vancomycin-like glycopeptide antibiotics and found two new biosynthetic gene clusters. In addition, an *in silico* examination from the Global Ocean Metagenomic Survey found 20 new lantibiotic cyclases.

There are challenges to metagenomic approaches to discover new biologically active molecules. Only 40% of enzymatic activities present in a sample can be expressed in *E. coli.*. In addition, the purification and isolation of eDNA is essential but difficult when the sources of obtained samples are poorly understood. However, collaborative efforts from individuals from diverse fields including bacterial genetics, molecular biology, genomics, bioinformatics, robots, synthetic biology, and chemistry can solve this problem together and potentially lead to the discovery of many important biologically active molecules.

Protein Phosphorylation

Posttranslational modification of proteins with phosphate groups has proven to be a key regulatory step throughout all biological systems. Phosphorylation events, either phosphorylation by protein kinases or dephosphorylation by phosphatases, result in protein activation or deactivation. These events have an immense impact on the regulation of physiological pathways, which makes the ability to dissect and study these pathways integral to understanding the details of cellular processes. There exist a number of challenges—namely the sheer size of the phosphoproteome, the fleeting nature of phosphorylation events and related physical limitations of classical biological and biochemical techniques—that have limited the advancement of knowledge in this area. A recent review provides a detailed examination of the impact of newly developed chemical approaches to dissecting and studying biological systems both in vitro and in vivo.

Through the use of a number of classes of small molecule modulators of protein kinases, chemical biologists have been able to gain a better understanding of the effects of protein phosphorylation. For example, nonselective and selective kinase inhibitors, such as a class of pyridinylimidazole compounds described by Wilson, et al., are potent inhibitors useful in the dissection of MAP kinase signaling pathways. These pyridinylimidazole compounds function by targeting the ATP binding pocket. Although this approach, as well as related approaches, with slight modifications, has proven effective in a number of cases, these compounds lack adequate specificity for more general applications. Another class of compounds, mechanism-based inhibitors, combines detailed knowledge of the chemical mechanism of kinase action with previously utilized inhibition motifs. For example, Parang, et al. describe the development of a "bisubstrate analog" that inhibits kinase action by binding both the conserved ATP binding pocket and a protein/peptide recognition site on the specific kinase. While there is no published in vivo data on compounds of this type, the structural data acquired from in vitro studies have expanded the current understanding of how a number of important kinases recognize target substrates. Interestingly, many research groups utilized ATP analogs as a chemical probe to study kinases and identify their substrates.

The development of novel chemical means of incorporating phosphomimetics into proteins has provided important insight into the effects of phosphorylation events. Historically, phosphorylation events have been studied by mutating an identified phosphorylation site (serine, threonine or tyrosine) to an amino acid, such as alanine, that cannot be phosphorylated. While this approach has been successful in some cases, mutations are permanent in vivo and can have potentially detrimental effects on protein folding and stability. Thus, chemical biologists have developed new ways of investigating protein phosphorylation. By installing phospho-serine, phospho-threonine or analogous phosphonate mimics into native proteins, researchers are able to perform in vivo studies to investigate the effects of phosphorylation by extending the amount of time a phosphorylation event occurs while minimizing the often-unfavorable effects of mutations. Protein semisynthesis, or more specifically expressed protein ligation (EPL), has proven to be successful techniques for synthetically producing proteins that contain phosphomimetic molecules at either the C- or the N-terminus. In addition, researchers have built upon an established technique in which one can insert an unnatural amino acid into a peptide sequence by charging synthetic tRNA that recognizes a nonsense codon with an unnatural amino acid. Recent developments indicate that this technique can also be employed in vivo, although, due to permeability issues, these in vivo experiments using phosphomimetic molecules have not yet been possible.

Advances in chemical biology have also improved upon classical techniques of imaging kinase action. For example, the development of peptide biosensors—peptides containing incorporated fluorophore molecules—allowed for improved temporal resolution in in vitro binding assays. Experimental limitations, however, prevent this technique from being effectively used in vivo. One of the most useful techniques to study kinase action is Fluorescence Resonance Energy Transfer (FRET). To utilize FRET for phosphorylation studies, fluorescent proteins are coupled to both a phosphoamino acid binding domain and a peptide that can by phosphorylated. Upon phosphorylation or dephosphorylation of a substrate peptide, a conformational change occurs that results in a change in fluorescence. FRET has also been used in tandem with Fluorescence Lifetime Imaging Microscopy (FLIM) or fluorescently conjugated antibodies and flow cytometry to provide a detailed, specific, quantitative results with excellent temporal and spatial resolution.

Through the augmentation of classical biochemical methods as well as the development of new tools and techniques, chemical biologists have improved accuracy and precision in the study of protein phosphorylation.

Metal Complexes in Medicine

Metal complexes have many characteristics that can be advantageous in drug design. In comparison to organic-based medicines, metal complexes have many more Coordination Numbers, geometries, and oxidation/reduction states that can be used to make structures that interact with targets in unique ways unavailable to most organic molecules. In addition, the cationic metal is advantageous in complexing with charged targets within biological systems like the phosphate backbone of DNA. Targets of metal-based medicines include DNA, proteins, and enzymes.

Metal Complexes Targeting DNA

DNA has been the primary target of metal complexes due to the ability of cationic metal interacting with the anionic backbone of DNA. The anticancer chemotherapy drug cisplatin covalently binds to DNA, which disrupts transcription and leads to programmed cell death. Assuming early detection, cisplatin cures almost all cases of testicular cancer. This drug, however, has severe side effects and great effort is being made to improve drug delivery including attachment to single-walled carbon nanotubes, encapsulation in proteins cages, among other clever strategies.

Another major effort for anticancer metal-based drugs centers around stabilization of the G-quadruplex of DNA. In general, these drugs have a non-covalent interaction with the G-quadruplex as well as a planar positively charged structure.

Metal Complexes Targeting Enzymes and Proteins

Though DNA has been a primary target for inorganic medicines, enzymes, and proteins also can be modulated through interactions with these compounds. Metal complexes can interact with the amino acids with the highest reduction potential (histidine, cysteine, and selenocysteine). Metals used in such complexes include gold, platinum, ruthenium, vanadium, cobalt and others. Several new potential therapeutic complexes are currently in the process of discovery and investigation.

Gold

Some gold complexes are showing potential as medicines. A rheumatoid arthritis drug (auranofin, a gold(I) phosphine complex) has shown value in treating parasitic disease through inhibiting thioredoxin glutathione reductase.

Platinum

Along with cisplatin, many other platinum complexes are potential therapeutics. Like auranofin, terpyridine platinum inhibits thioredoxin reductase with nanomolar IC50. This complex also is an inhibitor of the common target enzyme topoisomerase I. Yet another family of complexes with potential anticancer properties are dichloro(SMP)-platinum(II) complexes. These complexes target the matrix metalloproteinase, where the complex coordinates with amino acids of the enzyme in the coordination sites previously held by chlorides, and through the smp ligand. As seen by these few examples, platinum complexes are a particularly active area of research for metal-based medicines.

Ruthenium

Ruthenium complexes have anticancer activity. A library of glutathione transferase inhibitors were created through a combination of ethacrynic acid (a known inhibitor of the enzyme) and ruthenium complexes.

Vanadium

Vanadium complexes have been used in multiple therapeutic settings. A new area in which vanadium may have a great medicinal impact is through the oxovanadium porphyrin complexes. These complexes have demonstrated HIV-1 reverse transcriptase inhibition in vitro.

Issues and Outlook

Though there is currently much excitement in the field of metal-based medicines, many challenges still face researchers. One such challenge is selectivity of complexes in vivo. Many of these complexes can bind to common proteins like serum albumin in addition to other proteins with amino acids that are common in protein–metal complex interactions like histidine, cysteine, and sele-nocysteine. Along with selectivity issues, much is yet unknown about mechanisms through which metal complexes interact with proteins. How complexing between a given metal complex and tar-get protein or enzyme occurs is often unknown or unclear and requires much more elucidation before truly effective metal complexes can be designed and delivered. Currently, physicians utilize very few metal-based medicines in the clinics. For example, none of the 21 drugs approved by the U.S. Food and Drug Administration (FDA) in 2008 were inorganic. However, with the success of cisplatin in cancer treatment, it is not unreasonable to anticipate more metal complexes will be actively used in the treatment of diseases.

Synthetic Biology

Synthetic biology focuses on the manipulation of biological components to form new systems or the generation of living systems with synthetic parts. The canonical idea of synthetic biology is the creation of new life, but recently it has come to include bioengineering in terms of the use of in-terchangeable components to give novel outputs. In the search for modular parts, it is most facile if the building blocks contribute independently to the function of the whole unit so that the mod-ules can be recombined in predictable ways. It is useful for synthetic biologists to define "life": in this context, to be alive an organism must be capable of Darwinian evolution – genetic mutation, self-replication and inheritance of mutations.

Synthetic Cells

J. Craig Venter's group has created the first "synthetic" cell – the first cells to exist with fully syn-thetic DNA. Venter was able to manipulate the synthetic genome to dictate the proteins expressed in the organism. Note that these were not fully synthetic cells but that the synthetic DNA was able to take over all metabolic processes necessary for cell survival and proliferation.

DNA as Interchangeable Parts

DNA is composed of repeating modular units consisting of an anion phosphate group that forms the polyanion backbone, and nucleotide base pairs that engage in Watson-Crick base pairing to form the double strand. Because the molecular recognition of DNA is based mostly on the polyan-ion backbone, the nucleotides can be modified without altering the structural integrity of the DNA. Steven Benner's group has generated an artificial genetic alphabet of eight new base pairs that can be amplified by polymerase chain reaction; this indicates that these base pairs can be used in sys-tems that undergo Darwinian evolution.

Proteins as Interchangeable Parts

Amino Acids

Amino acids are poor modular building-blocks because they do not act independently and there is a fundamental lack of understanding about the relationship between linear amino acid sequences and the folding and functionality of proteins. Chemical biologists have been able to model, design, and synthesize peptides and evaluate their function.

Protein Secondary Structure

Modules consisting of protein secondary structure can be designed to perform specific functions; for example, it has been demonstrated that alpha helices can be used as functional peptide catalysts. The Ghadiri group has created a template peptide that promotes the ligation of two modified helices by bringing the helices into close proximity by specifically designed hydrophobic interactions of the helices with the template.

Folded Proteins

Fully folded proteins can be combined in novel ways to generate specific non-natural outcomes. This is highly useful commercially from drug development to the production of polymers – one can imagine the economic benefits if scientists can design systems in which proteins catalyze reactions without the necessity of excessive human intervention to produce commercially relevant materials. For example, the Keasling group has developed a series of proteins that catalyze conversion of acetyl CoA, a common cellular metabolite, into a precursor for the potent antimalarial drug artemisinin.

Modifying Molecular Switches

Signaling pathways can be modified to be turned on or off by non-natural ligands or inputs to the system. For instance, systems can be modified so that they are autoinhibited by non-natural proteins that release their inhibition upon binding with a specific molecule that is different from the natural signaling molecule of the path. This allows new approaches to studying signal circuits specifically and with user-designed inputs.

Chemical Approaches to Stem-cell Biology

Advances in stem-cell biology have typically been driven by discoveries in molecular biology and genetics. These have included optimization of culture conditions for the maintenance and differentiation of pluripotent and multipotent stem-cells and the deciphering of signaling circuits that control stem-cell fate. However, chemical approaches to stem-cell biology have recently received increased attention due to the identification of several small molecules capable of modulating stem-cell fate in vitro. A small molecule approach offers particular advantages over traditional methods in that it allows a high degree of temporal control, since compounds can be added or removed at will, and tandem inhibition/activation of multiple cellular targets.

Small molecules that modulate stem-cell behavior are commonly identified in high-throughput screens. Libraries of compounds are screened for the induction of a desired phenotypic change in cultured stem-cells. This is usually observed through activation or repression of a fluorescent reporter or by detection of specific cell surface markers by FACS or immunohistochemistry. Hits are then structurally optimized for activity by the synthesis and screening of secondary libraries. The cellular targets of the small molecule can then be identified by affinity chromatography, mass spectrometry, or DNA microarray.

A trademark of pluripotent stem-cells, such as embryonic stem-cells (ESCs), is the ability to self-renew indefinitely. The conventional use of feeder cells and various exogenous growth factors in the culture of ESCs presents a problem in that the resulting highly variable culture conditions make the long-term expansion of un-differentiated ESCs challenging. Ideally, chemically defined culture conditions could be developed to maintain ESCs in a pluripotent state indefinitely. Toward this goal, the Schultz and Ding labs at the Scripps Research Institute identified a small molecule that can preserve the long-term self-renewal of ESCs in the absence of feeder cells and other exogenous growth factors. This novel molecule, called pluripotin, was found to simultaneously inhibit multiple differentiation inducing pathways.

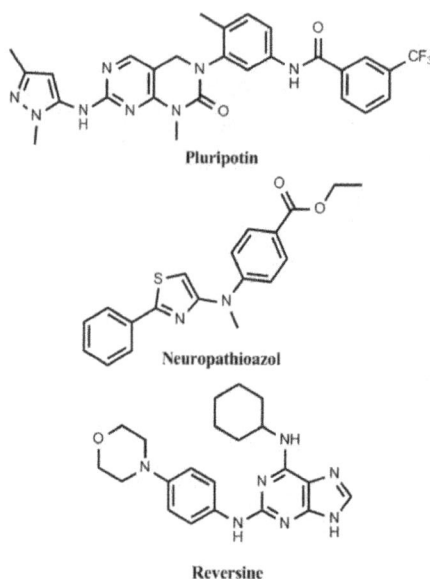

Small molecule modulators of stem-cell fate.

The utility of stem-cells is in their ability to differentiate into all cell types that make up an organism. Differentiation can be achieved in vitro by favoring development toward a particular cell type through the addition of lineage specific growth factors, but this process is typically non-specific and generates low yields of the desired phenotype. Alternatively, inducing differentiation by small molecules is advantageous in that it allows for the development of completely chemically defined conditions for the generation of one specific cell type. A small molecule, neuropathiazol, has been identified which can specifically direct differentiation of multipotent neural stem cells into neurons. Neuropathiazol is so potent that neurons develop even in conditions that normally favor the formation of glial cells, a powerful demonstration of controlling differentiation by chemical means.

Because of the ethical issues surrounding ESC research, the generation of pluripotent cells by reprogramming existing somatic cells into a more "stem-like" state is a promising alternative to the use of standard ESCs. By genetic approaches, this has recently been achieved in the creation of ESCs by somatic cell nuclear transfer and the generation of induced pluripotent stem-cells by viral transduction of specific genes. From a therapeutic perspective, reprogramming by chemical means would be safer than genetic methods because induced stem-cells would be free of potentially dangerous transgenes. Several examples of small molecules that can de-differentiate somatic cells have been identified. In one report, lineage-committed myoblasts were treated with a compound, named reversine, and observed to revert to a more stem-like phenotype. These cells were then shown to be capable of differentiating into osteoblasts and adipocytes under appropriate conditions.

Stem-cell therapies are currently the most promising treatment for many degenerative diseases. Chemical approaches to stem-cell biology support the development of cell-based therapies by enhancing stem-cell growth, maintenance, and differentiation in vitro. Small molecules that have been shown to modulate stem-cell fate are potential therapeutic candidates and provide a natural lean-in to pre-clinical drug development. Small molecule drugs could promote endogenous stem-cells to differentiate, replacing previously damaged tissues and thereby enhancing the body's own regenerative ability. Further investigation of molecules that modulate stem-cell behavior will only unveil new therapeutic targets.

Fluorescence for Assessing Protein Location and Function

Fluorophores and Techniques to Tag Proteins

Organisms are composed of cells that, in turn, are composed of macromolecules, e.g. proteins, ribosomes, etc. These macromolecules interact with each other, changing their concentration and suffering chemical modifications. The main goal of many biologists is to understand these interactions, using MRI, ESR, electrochemistry, and fluorescence among others. The advantages of fluorescence reside in its high sensitivity, non-invasiveness, safe detection, and ability to modulate the fluorescence signal. Fluorescence was observed mainly from small organic dyes attached to antibodies to the protein of interest. Later, fluorophores could directly recognize organelles, nucleic acids, and important ions in living cells. In the past decade, the discovery of green fluorescent protein (GFP), by Roger Y. Tsien, hybrid system and quantum dots have enable assessing protein location and function more precisely. Three main types of fluorophores are used: small organic dyes, green fluorescent proteins, and quantum dots. Small organic dyes usually are less than 1 kD, and have been modified to increase photostability, enhance brightness, and reduce self-quenching. Quantum dots have very sharp wavelength, high molar absorptivity and quantum yield. Both organic dyes and quantum dyes do not have the ability to recognize the protein of interest without the aid of antibodies, hence they must use immunolabeling. Since the size of the fluorophore-targeting complex typically exceeds 200 kD, it might interfere with multiprotein recognition in protein complexes, and other methods should be use in parallel. An advantage includes diversity of properties and a limitation is the ability of targeting in live cells. Green fluorescent proteins are genetically encoded and can be covalently fused to your protein of interest. A more developed genetic tagging technique is the tetracysteine biarsenical system, which requires modification of the targeted sequence that includes four cysteines, which binds membrane-permeable biarsenical molecules, the green and the red dyes "FlAsH" and "ReAsH", with picomolar affinity. Both fluorescent proteins and biarsenical tetracysteine can be expressed in live cells, but present major limitations in ectopic expression and might cause lose of function. Giepmans shows parallel applications of targeting methods and fluorophores using GFP and tetracysteine with ReAsH for α-tubulin and β-actin, respectively. After fixation, cells were immunolabeled for the Golgi matrix with QD and for the mitochondrial enzyme cytochrome with Cy5.

Protein Dynamics

Fluorescent techniques have been used assess a number of protein dynamics including protein tracking, conformational changes, protein–protein interactions, protein synthesis and turnover, and enzyme activity, among others.

Three general approaches for measuring protein net redistribution and diffusion are single-particle tracking, correlation spectroscopy and photomarking methods. In single-particle tracking, the individual molecule must be both bright and sparse enough to be tracked from one video to the other. Correlation spectroscopy analyzes the intensity fluctuations resulting from migration of fluorescent objects into and out of a small volume at the focus of a laser. In photomarking, a fluorescent protein can be dequenched in a subcellular area with the use of intense local illumination and the fate of the marked molecule can be imaged directly. Michalet and coworkers used quantum dots for single-particle tracking using biotin-quantum dots in HeLa cells.

One of the best ways to detect conformational changes in proteins is to sandwich said protein between two fluorophores. FRET will respond to internal conformational changes result from re-orientation of the fluorophore with respect to the other. Dumbrepatil sandwiched an estrogen receptor between a CFP (cyan fluorescent protein) and a YFP (yellow fluorescent protein) to study conformational changes of the receptor upon binding of a ligand.

Fluorophores of different colors can be applied to detect their respective antigens within the cell. If antigens are located close enough to each other, they will appear colocalized and this phenomenon is known as colocalization. Specialized computer software, such as CoLocalizer Pro, can be used to confirm and characterize the degree of colocalization.

FRET can detect dynamic protein–protein interaction in live cells providing the fluorophores get close enough. Galperin *et al.* used three fluorescent proteins to study multiprotein interactions in live cells.

Tetracysteine biarsenical systems can be used to study protein synthesis and turnover, which requires discrimination of old copies from new copies. In principle, a tetracysteine-tagged protein is labeled with FlAsH for a short time, leaving green labeled proteins. The protein synthesis is then carried out in the presence of ReAsH, labeling the new proteins as red.

One can also use fluorescence to see endogenous enzyme activity, typically by using a quenched activity based proteomics (qABP). Covalent binding of a qABP to the active site of the targeted enzyme will provide direct evidence concerning if the enzyme is responsible for the signal upon release of the quencher and regain of fluorescence.

The unique combination of high spatial and temporal resolution, nondestructive compatibility with living cells and organisms, and molecular specificity insure that fluorescence techniques will remain central in the analysis of protein networks and systems biology.

Applications of DNA Microarrays in Chemical Biology

Planar surfaces functionalized with single- or double-strand nucleic acids have enabled researchers to address a variety of salient biological and biochemical questions in recent years. The general architecture of modern DNA microarrays reflects the historical progression from the sequence-specific probing of whole chromosomes immobilized on glass slides (as early as 1961 with fluorescent *in situ* hybridization) and the low-density porous membrane arrays available since the early 1990s, to the high-density (10^2-10^4 features/mm^2) solid support platforms that exist today. The massively parallel processing capabilities of these picomolar-range contemporary arrays provide for the generation of large data sets and multiplexed analysis. Furthermore, several top-down and bottom-up assembly methodologies provide researchers with the option for "in-house" production of arrays from custom oligonucleotide libraries or the use of commercial genome chips, notably those developed by Affymetrix and Agilent Technologies.

DNA microarrays can be used to conduct several general types of experiments, most of which relying on the hybridization of fluorescently labeled single-strand DNA molecules isolated from a biological sample to their single-strand complement probes presented on an array. One of the earliest conceived applications for DNA microarrays was for single-nucleotide polymorphism (SNP) genotyping. Since SNPs are a "quick and dirty" approach to detect genetic indicators of pathologies and lineages, arrays in theory provide a facile method for diagnosis; this was confirmed experimentally in the late 1990s in the successful SNP analysis of human tumors. Although there are currently commercially available arrays (e.g. bovine mapping chips) to characterize SNPs, it seems likely that the nascent availability of high-throughput and low-cost pyrosequencing will become the preferred method of recognition, or replace the need for SNP detection altogether with rapid whole-genome sequencing.

A different application of microarray technology that has become the gold standard for RNA analysis in recent years is the widespread utilization of expression microarrays, or "gene chips". Gene chip preparation calls for the quantitative reverse transcription of the total cellular RNA pool into labeled and fragmented single-strand DNA prior to hybridization-based capture. Up- and down-regulation of genes in response to stressors or disease states are quantitatively compared in cell lines and organisms. Coupled expression microarray and quantitative proteomics experiments have allowed for the in-depth exploration of the oftentimes non-linear relationship between the abundance of a particular transcribed message and that of its corresponding translated protein. These integrative studies, partially enabled by quantitative DNA microarray technology, have been successfully applied to a variety of biological systems, including yeast, bovine, mouse, bacterial, and human. The expression analysis community has amassed such a significant amount of expression microarray data that they are freely available in public databases.

These types of surfaces can also be used to analyze DNA-protein interactions on a genome-wide scale via chromatin immunoprecipitation, followed by an array-based analysis of the DNA (ChIP-chip). ChIP-chip experiments are enabled by the co-purification of a DNA-binding protein of interest with its corresponding genomic loci when a cross-linked chromatin extract is probed with an antibody to said protein. After purification, amplification and labeling, the DNA is applied to a microarray representing the entire genome; the data are plotted as a histogram that resolves the specific genomic regions associated with that protein. ChIP-chip experiments have provided the scientific community with a wealth of information about the steady-state genomic locations of DNA-binding proteins, such as histones, transcription factors, and polymerase machinery, and have also been successfully applied to studies on the dynamics of transcription factor binding. The data from these experiments may be further manipulated to computationally derive consensus binding sequences for some transcription factors, giving the opportunity for insight into the in vivo behavior of the factor, deeper than simple information about localization.

DNA microarrays are also amenable to the direct analysis of protein–DNA interactions in kinetic binding assays as analyzed by surface plasmon resonance (SPR). This experimental approach also relies on single-strand DNA immobilized on a high-density array; however, the quantitative readout is based on a change in the optical properties of the DNA-functionalized surface when a protein flowed over the surface binds to the sequence in a particular surface feature. DNA-functionalized arrays analyzed with SPR in this way have yielded kinetic data regarding fundamental molecular biological processes. Recently, SPR analysis of a DNA microarray and components of the DNA replication machinery helped to elucidate the biochemical nuances of the replication fork.

High-density DNA microarrays have emerged as an important component of the chemical biology toolkit. The existing technology allows for the construction of customizable, as well as general, arrays and provides researchers with the opportunity to generate robust data from many different types of biological inputs. Considering the relatively recent shift in the scientific community away from binary perturbation/readout studies and toward "big science" and large data sets, it seems likely that DNA microarrays will continue to enable pertinent biological research for many years to come.

Applications of Chemical Biology in Drug Discovery

Chemical biology approaches can help answer important questions of relevance to small molecule drug discovery projects. This includes questions related to the characterization of protein targets and the molecular pharmacology of small molecule drugs that modulate target function.

Chemical Biology Approaches to Characterize Protein Targets

Protein Target Characterization		
Question	Why ask?	Chemical Biology Technique
What is my target?	Enables target-based drug discovery on hits/mechanisms from phenotypic screens.	Affinity chemoproteomics; Mutagenesis; Phenotypic screening; Chemogenomics; Proteomics
What is the subcellular distribution of my target?	Location of target may influence screening assays or inhibitor design; active and inactive species of the target may localize to different cellular regions; target may be activated by particular environment owing to localization to a particular organelle (for example, acidic lysosome)	Imaging; Microscopy
Does my target exist in multiple forms, and does this vary across tissues and species?	May reveal species differences, splice variants, relevance of full length target vs. catalytic domains for screening; splice variants of the target may have different protein domains, activity, cellular location, tissue distribution, and affinity for substrate; knowing correct sequence cDNA enables potential to express recombinant protein or to generate overexpression cell line which can inform choice of primary assay and screening sequence.	Computational biology; Genotype-tissue-expression analysis; Proteomics
What is the endogenous ligand for my target and its concentration in the diseased state?	Allows one to understand what biological pathways are being modulated (one of the keys for establishing a physiologically relevant assay); allows one to theorize what will happen if endogenous substrate levels are increased by inhibiting the desired target.	Immunoprecipitation; Metabolomics; Peptide Microarrays; Peptidomics
How well characterized is the interaction of the endogenous substrate with my target?	Informs chemical feasibility, screening strategy and medicinal chemistry approach (for example, substrate concentration and Km for enzyme target)	Biochemical enzyme and cellular activity assays

Is my target post-translationally modified?	Can rationalize differences in affinity/efficacy between biochemical in vitro and cellular/in vivo systems; informs primary assay choice and screen sequence.	Chemical Probe; Immunodetection; Metabolomics; Proteomics; Phosphoproteomics
What other proteins are influential in regulating substrate concentrations?	Can represent alternate strategies for modulating the desired pathway. Can lead to complimentary pharmacology.	Computational Biology; RNAi
What is the turnover of my target and is this affected by my compound?	Especially important to understand the cellular efficiencies of covalent modalities.	Chemical probe; Immunoprecipitation; Pulse/chase; SILAC/MS
What is the abundance of my target, and does it vary?	Abundance of a protein might vary by tissue, disease state, or as a response to drug action.	Targeted quantitative proteomics coupled with immunocapture
Does my target interact with other proteins and what is the consequence of these interactions?	Can lead to a better understanding of signaling pathway; informs screening assay; could be important that protein forms hetero or homodimers.	Immunoprecipitation
What potential off-targets are most closely related to my target sequence and function?	Informs screening sequence design; important to consider not only targets that are closely related in terms of binding site sequence, but also those most closely related in a chemogenomic sense; understand which tissues express off-targets.	Computational Biology

Protein Target Characterization

- What is my target?

  - *Affinity chemoproteomics* can be used to identify the targets of compounds that have shown activity in a phenotypic or pathway screen. For example, this approach was used to identify of MTH1 as a potential anticancer target of the (S) form of Crizotinib. Affinity chemoproteomics technology was also used to identify the BET bromodomains as the molecular targets of a series of small molecule modulators of Apolipoprotein A1 (ApoA1).

  - *Mutagensis* is important target validation approach is to generate a catalytically-dead target protein to tease apart the role of the enzymatic and scaffolding functions. This method is often used in kinase validation experiments where the pharmacological performance of a putative inhibitor is tested in cells with the kinase knocked out and replaced through transfection of the recombinant wild-type protein or its kinase dead mutant. This approach was used recently to confirm DCLK1 as the functional anti-proliferative target of a previously developed LRRK2 inhibitor for Parkinson's disease

- Does my target exist in multiple related forms and does this vary across different tissues and species?

  - *Genotype-tissue expression project (GTEx)* has generated a database of mRNA expression levels across multiple tissues. An illustration of the use of this resource shows that PDE4B is expressed in multiple forms that vary across human tissues. Being aware of these different forms has enabled development of a selective inhibitor targeting PDE4B in the brain through taking advantage of differences in sequences in the long and short forms outside the active site.

- What is the endogenous ligand for my target and its concentration in the diseased state?

  - *Metabolomics* in combination with clickable chemical reporters can enable chemoproteomic methods to be applied to substrate identification, as shown recently for protein lipidation.

- Is my target post-translationally modified?

  - Dimedone *chemical probes* that chemoselectively react with sulfenic acid residues in EGFR have shown how cysteine oxidation can be use for enzymatic regulation.

- What is the turnover of my target and is this affected by my compound?

  - *Chemical probes* of BTK were used to confirm its slow turnover rate.

  - *Immunoprecipitation* of EGFR has been used to confirm increased protein half-livey, likely due to a concomitant decrease in binding to the Cbl ubiquitin ligase responsible for regulating the degradation of EGFR. This increased half-life could result in a need to change the projected dose or frequency. In addition to mutations altering the turnover rate of a protein, small molecule drugs can also change the turnover and thus the amount of protein in the cell. For instance, many kinases are known to be clients of the Hsp90-Cdc37 chaperone system and recent studies have found that some ATP-competitive inhibitors disrupt the ability of Cdc37 to bind to the target kinase and recruit it to Hsp9015.

- What potential off- targets are most closely related to my target sequence and function?

  - *Computational biology* and structural bioinformatic analyses of protein sequences and binding pocket topology can be used to delineate which proteins are similar to a desired target's biological sequence as well as its binding site shape. Additionally, chemoinformatic approaches using in silico tools to predict potential off-targets based on the structural similarity between a novel small molecule and previously known compounds with different pharmacology are also useful.

Chemical Biology Approaches to Characterize Molecular Pharmacology

Molecular Pharmacology Characterization		
Question	Why ask?	Chemical Biology Technique
What target(s) and off-targets does my molecule bind?	Key to understanding what drives molecule efficacy and potential safety liabilities; allows team to understand mechanism of action, pathways affected and target that drives efficacy; enables target-based drug discovery and understanding or improvement of potency and selectivity.	Activity probes; Affinity capture; ABPP; DARTS; Photoaffinity; Proteomics; Protein microarrays; SPROX, Thermal aggregation; Y3H assays
Where does my molecule bind?	Characterization of binding kinetics and binding site may be key to functional translation and pharmacokinetic-pharmacodynamic relationships.	Hydrogen-deuterium exchange
How does my molecule bind?	Characterization of binding kinetics and binding site may be key to functional translation and PK/PD.	Kinetics

What is the tissue distribution of my molecule?	Assuming plasma exposure reflects target tissue exposure is often incorrect. Affects Pillar 1.	MALDI MS; PK: Protein binding; Tissue distribution
What are the functional consequences to my target when my molecule binds?	Provides an understanding of how the molecule works; binding could modulate target degradation, stabilization, translocation or interactions with other proteins.	Immunodetection; Proteomics
How much target occupancy do I need to drive my relevant biological phenotype?	Allows team to develop Pillar II confidence and link to Pillar III. Key enabler to help define Ceff.	Occupancy probes; "Three pillars in a tube"

Molecular Pharmacology Characterization

- What target(s) and off-targets does my molecule bind?

  - *Activity probes* such as nucleotide acyl phosphates (KiNativ™) probes that react chemoselectively with the catalytic lysine of >80% of all known kinases can be used to determine many of the kinase targets of a molecule.

  - *ABPP* using probes derived from covalent kinase inhibitors, have been used to determine the cellular targets of these covalent inhibitors in living cells.

  - *Chemoproteomics* was used to assess the multiple binding partners of HDAC inhibitors in protein complexes scaffolded by ELM-SANT domain subunits. *Thermal proteomics*, where one looks for proteins that show increased thermal stability at elevated temperatures due to compound binding, has been used to identify off-targets of several kinase inhibitors including the BRAF inhibitor Vemurafenib. *Affinity capture* where a chemical probe is tethered to a solid support, was used to identify cereblon (CRBN) as a thalidomide binding protein.

  - *DARTS (Drug Affinity Responsive Target Stability);* takes advantage of a reduction in the protease susceptibility of the target protein upon drug binding. This technique was used to confirm binding biological binding partners for a number of molecules: rapamycin and FKBP with FKBP12 and resveratrol with eIF4A.

  - *SPROX (Stability of Proteins from Rates of Oxidation);* utilizes the chemical denaturant-dependent oxidation rates of methionine residues in a protein to report on the thermodynamic properties of its global and/or subglobal unfolding/refolding reactions.

- What is the tissue distribution of my molecule?

  - Quantitative *mass spectrometry (MS)* was recently used to determine the concentrations of the chemotherapeutic agent YM155 in various cancer cells.

- What are the functional consequences to my target when my molecule binds?

  - The use of shRNA gene silencing combined with *immunoblotting* showed that BRAF inhibitors GDC-0879 and PLX4720 block MAPK signaling in BRAF (V600E) tumors while these same compounds activate the RAF-MEK-ERK pathway in KRAS mutant and RAS/RAF wild-type tumors.

- How much target occupancy do I need to drive my relevant biological phenotype?

  - A clickable covalent *occupancy probe* of fatty acid amide hydrolase (FAAH) has been used to relate target engagement to anandamide elevation and efficacy in models of rodent inflammatory and neuropathic pain. Target occupancy quantification can also be used to validate the relevance of a drug target identified from phenotypic screening. A clickable covalent probe of the mRNA decapping enzyme DcpS, which used a sulfonyl fluoride warhead to target a reactive tyrosine in the binding site of the enzyme, enabled the assessment of DcpS target engagement of a diaminoquinazoline inhibitor (developed from a phenotypic screen for the treatment of spinal muscular atrophy).

Goal of Chemical Biology

- In the postgenomic era it is a major challenge to determine how proteins function together in complex molecular systems to drive a diverse set of cellular processes. Molecularly focused technologies that can probe and manipulate protein function in a controlled manner can manage this great job. Chemical Biology at its position at the interface between chemistry and biology can provide a unique platform for the development of such a molecular toolkit.

- Chemical Biology research is diverse. Main goal of Chemical Biology is to synthesize molecules that can be used as tools to selectively and reversibly modulate proteins.

- The synthesis of molecules to study extra- and intracellular signaling is another aim of Chemical Biology as well as some aspects of Glycobiology which is concerned with the study of different types of sugar molecules in the cell. In Chemical Biology studies, synthetic variations of sugar molecules can be used as tools for research.

- Another important branch of Chemical Biology uses endogenous biomolecules to develop chemical processes and/or materials.

- Research in the field is often concerned with the understanding of biological functions in the healthy individual as well as of the pathological mechanisms related to disease conditions including cancer, neurodegenerative disorders, renal and pulmonary dysfunctions and metabolic disorders.

- The field is therefore important for the generation of knowledge and tools for basic science as well as for the study of disease mechanisms. For many diseases, it is also important for the production of countermeasures and preventive actions.

- Small Molecule Approach: Building of library of bioactive small molecules is an example of Chemical Biology research. Recent research of small molecule approach to Chemical Biology includes:

  o The study of the different subtypes of Adenosine receptors reported by Jacobson et al., and their agonists and antagonists (activators and blockers).

  o Adenosine receptors are known or suspected to be involved in multiple diseases and conditions including Inflammation, Endocrine disorders, Cancer, Vision disorders, Renal disorders, Pulmonary disorders, Dementia, Anxiety, Pain, Parkinson's disease, Sleep disorders and Ischaemia. It is therefore clear that molecular tools to reversibly and selectively manipulate the functions of different subtypes of receptors involved in the conditions may have the potential to become useful tools in the research and also in some cases useful therapeutics.

o   Research also emphasized on the manipulation of receptors and reengineer the binding sites on the receptors is the target of chemical biologists. This type of approach could lead to insights into the accuracy of G-protein coupled receptor modeling, signaling pathways, and not least, the design of small molecules to be able to rescue disease-related mutations and do small-molecule directed gene therapy.

o   The combination of tailoring of small molecules and their protein targets is therefore of special interest.

o   Naturally occurring small molecules called natural products or their derivatives encampus a substantial fraction of the current pharmacopeia. Understanding the relationship of protein targets of natural products to heritable disease genes by comparing the biological functional connections between genes and gene products, e.g., protein/protein interactions (network connectivities) of these targets and genes will lead to the drug discovery for disease treatment at protein level. By determining whether natural products are intrinsically suited for targeting disease genes and whether their enrichment among current drugs reflects a historical focus or special properties intrinsic to these molecules. Prof. Stuart L. Schreiber is working in a data-driven way, to discover whether or not the propensity of natural products for interaction with biological targets is an advantage for probe or drug discovery directed at the genes determined to be causal for human disease.

o   They are successful in identifying natural product targets and evaluating a database of natural products and targets from GVKBio. They have standardized 5581 target names and species to human proteins, as either direct natural product targets or orthologous human target proteins, and mapped these targets to 946 human proteins with connections in STRING. For human disease genes, they combined 3655 genes contained in the OMIM Morbid Map with 1580 genes from a genome-wide association study SNP database and mapped these to 2681 human proteins with connections in STRING.

•   Schreiber's results indicate that targets of natural products are highly connected, more than genes connected to human disease. This finding indicates that targeting at protein level is much more beneficial than at gene level for diagnostic.

•   Many natural products act as basic defense mechanisms against invaders in the absence of tissue specialization or an advanced immune response leading to the death of the invading organism. Therefore, the highly connected proteins can be targeted by natural products which will interrupt the activities of essential protein of the invader. Therefore the ultimate goal of chemical biology remains in the development of library of small natural molecules to target the root cause of the diseases and thus the treatment of all the heritable diseases can better way be done.

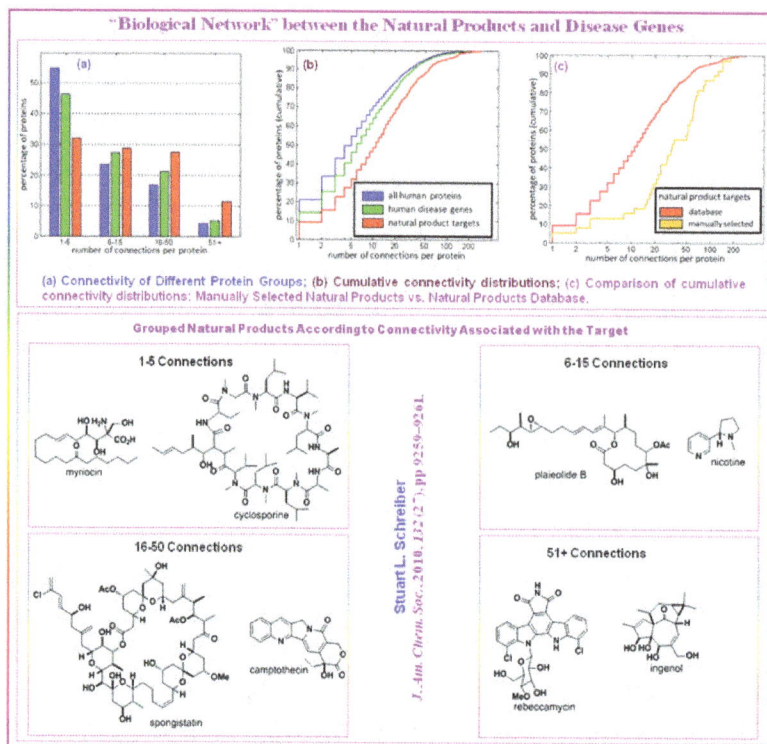

The small molecular approach of Chemical biology.

Manipulation of Biological Activity with Small Molecule:

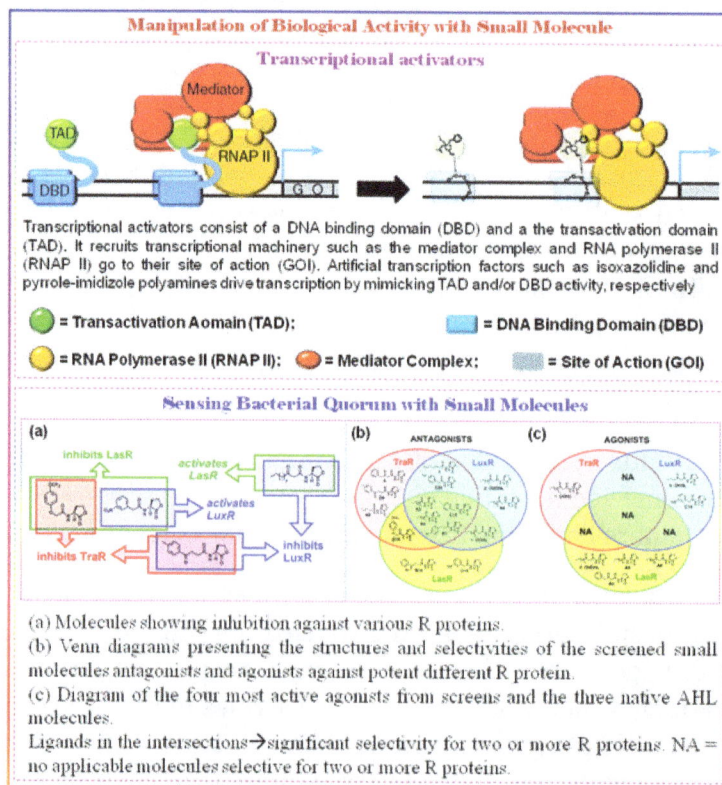

Manipulation of biological activity with small molecules.

In the postgenomic era it is major challenge to determine how proteins function together in complex molecular systems to drive a diverse set of cellular processes. Molecularly focused technologies that can probe and manipulate protein function in a controlled manner can manage this great job. Chemical Biology at its position at the interface between chemistry and biology can provide a unique platform for the development of such a molecular toolkit.

Due to the fundamental importance of transcriptional regulation in biological systems, small-molecule mimetic that promote exogenous control over transcription represent extremely valuable tools. Such study was reported by Jin Zhang et al., andHelen E. Blackwell et al.

Recent Advancement and Future Prospect of Chemical Biology:

Recent Advancement of Chemical Biology through Several Approaches

1. **Directed Development of Transition State (TS) Analogs:**
- Geometric and electrostatic potential mapping of enzymatic transition states by the use of several possible techniques of Chemistry, biology and computaion will allow us design transition state analogues molecules which will provide a wealth of enezyme inbitors drugs for arresting the enezyme related with diseases. Several of such inhibitors are in clinical trials or in preclinical development methods. Thus designing small molecules inhibitors is the fruitful approach for the drug development.

2. **A FRET-Based Strategy to Monitor Kinases Activity:**
- The activity assays on kinases based on FRET reporters as was reported by Allen, M. D. may find immediate application in high-throughput screening of small molecules drugs. It will also help in probing of multiple physiological and pharmacological events at subcellular locations in living cells in chemical and functional genomics studies. Thus, bridging of high-throughput technology with dynamic live-cell activity measurement can laid a establishment of mechanistic studies and versatile drug discovery processes targeting protein kinases.

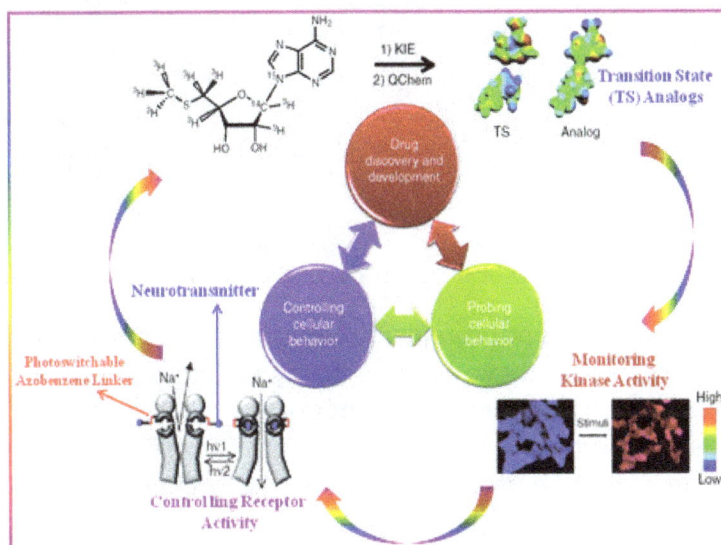

3. **Photoswitchable Gate: Remote Control of Receptor Activity:**
- Neurons have voltage-gated, ligands-gated, and temperature-gated ion channels but not light-gated ion channel. Recently a structure-based design of a new chemical gate was developed by Isacoff *et al.* that present light sensitivity to an ion channel. The gate includes a functional group for selective binding of a Shaker K$^+$ ion channel, a pore blocker and a photoisomerizable azobenzene linker. Light induced *Trans-Cis* isomerization of azobenzene linker allowed the photoswitchable gate to switch potential on and off. This strategy was used to control the activity of the Shaker potassium ion channel and ionotropic neurotransmitter receptors in a variety of systems, including zebrafish. The Advancement of such photoswitchable ion channel gate will allow allow rapid, precise and reversible control over neuronal firing. These will find potential applications for controlling the activity of such receptors and designing the potent molecules for future drug design targeting receptor.

Recent advancement of chemical biology.

The studies reported till the date accentuate the great impact of chemical biology to the study and modulate the biology and future scope of research at the interface of chemistry and biology.

Exploring the many existing small biological compounds and designing several enzyme inhibitors will also provide essential mechanistic probes and new candidates for drug discovery. Thus, many of such molecules showed and will continue to show the capability of targeting to an entire proteome. Also, the development of robust chemical/bio-chemical tools to probe and to investigate cellular principles at molecular levels will be of great achievement toward the success of conceptual approach, the Chemical Biology and thus, will facilitate the development of new therapeutics for treatment of diseases.

Proteomics

Robotic preparation of MALDI mass spectrometry samples on a sample carrier.

Proteomics is the large-scale study of proteins. Proteins are vital parts of living organisms, with many functions. The term *proteomics* was coined in 1997 in analogy with genomics, the study of the genome. The word *proteome* is a portmanteau of *prote*in and gen*ome*, and was coined by Marc Wilkins in 1994 while a PhD student at Macquarie University. Macquarie University also founded the first dedicated proteomics laboratory in 1995 (the Australian Proteome Analysis Facility – APAF).

The proteome is the entire set of proteins, produced or modified by an organism or system. This varies with time and distinct requirements, or stresses, that a cell or organism undergoes. Proteomics is an interdisciplinary domain that has benefited greatly from the genetic information of the Human Genome Project; it is also emerging scientific research and exploration of proteomes from the overall level of intracellular protein composition, structure, and its own unique activity patterns. It is an important component of functional genomics.

While *proteomics* generally refers to the large-scale experimental analysis of proteins, it is often specifically used for protein purification and mass spectrometry.

Complexity of the Problem

After genomics and transcriptomics, proteomics is the next step in the study of biological systems. It is more complicated than genomics because an organism's genome is more or less constant, whereas the proteome differs from cell to cell and from time to time. Distinct genes are expressed in different cell types, which means that even the basic set of proteins that are produced in a cell needs to be identified.

In the past this phenomenon was done by RNA analysis, but it was found not to correlate with protein content. It is now known that mRNA is not always translated into protein, and the amount of protein produced for a given amount of mRNA depends on the gene it is transcribed from and on the current physiological state of the cell. Proteomics confirms the presence of the protein and provides a direct measure of the quantity present.

Post-translational Modifications

Not only does the translation from mRNA cause differences, but many proteins are also subjected to a wide variety of chemical modifications after translation. Many of these post-translational modifications are critical to the protein's function.

Phosphorylation

One such modification is phosphorylation, which happens to many enzymes and structural proteins in the process of cell signaling. The addition of a phosphate to particular amino acids—most commonly serine and threonine mediated by serine/threonine kinases, or more rarely tyrosine mediated by tyrosine kinases—causes a protein to become a target for binding or interacting with a distinct set of other proteins that recognize the phosphorylated domain.

Because protein phosphorylation is one of the most-studied protein modifications, many "proteomic" efforts are geared to determining the set of phosphorylated proteins in a particular cell or tissue-type under particular circumstances. This alerts the scientist to the signaling pathways that may be active in that instance.

Ubiquitination

Ubiquitin is a small protein that can be affixed to certain protein substrates by enzymes called E3 ubiquitin ligases. Determining which proteins are poly-ubiquitinated helps understand how protein pathways are regulated. This is, therefore, an additional legitimate "proteomic" study. Similarly, once a researcher determines which substrates are ubiquitinated by each ligase, determining the set of ligases expressed in a particular cell type is helpful.

Additional Modifications

In addition to phosphorylation and ubiquitination, proteins can be subjected to (among others) methylation, acetylation, glycosylation, oxidation and nitrosylation. Some proteins undergo all these modifications, often in time-dependent combinations. This illustrates the potential complexity of studying protein structure and function.

Distinct Proteins are Made Under Distinct Settings

A cell may make different sets of proteins at different times or under different conditions, for example during development, cellular differentiation, cell cycle, or carcinogenesis. Further increasing proteome complexity, as mentioned, most proteins can undergo a wide range of post-translational modifications.

Therefore, a "proteomics" study can quickly become complex, even if the topic of study is restricted. In more ambitious settings, such as when a biomarker for a specific cancer subtype is sought, the proteomics scientist might elect to study multiple blood serum samples from multiple cancer patients to minimise confounding factors and account for experimental noise. Thus, complicated experimental designs are sometimes necessary to account for the dynamic complexity of the proteome.

Limitations of Genomics and Proteomics Studies

Proteomics gives a different level of understanding than genomics for many reasons:

- the level of transcription of a gene gives only a rough estimate of its *level of translation* into a protein. An mRNA produced in abundance may be degraded rapidly or translated inefficiently, resulting in a small amount of protein.

- as mentioned above many proteins experience *post-translational modifications* that profoundly affect their activities; for example some proteins are not active until they become phosphorylated. Methods such as phosphoproteomics and glycoproteomics are used to study post-translational modifications.

- many transcripts give rise to more than one protein, through alternative splicing or alternative post-translational modifications.

- many proteins form complexes with other proteins or RNA molecules, and only function in the presence of these other molecules.

- protein degradation rate plays an important role in protein content.

Reproducibility. One major factor affecting reproducibility in proteomics experiments is the simultaneous elution of many more peptides than can be measured by mass spectrometers. This causes stochastic differences between experiments due to data-dependant acquisition of tryptic peptides. Although early large-scale shotgun proteomics analyses showed considerable variability between laboratories, presumably due in part to technical and experimental differences between labs, reproducibility has been improved in more recent mass spectrometry analysis, particularly on the protein level and using Orbitrap mass spectrometers. Notably, targeted proteomics shows increased reproducibility and repeatability compared with shotgun methods, although at the expense of data density and effectiveness.

Methods of Studying Proteins

In proteomics, there are multiple methods to study proteins. Generally, proteins can either be detected using antibodies (immunoassays) or using mass spectrometry. If a complex biological sample is analyzed, either a very specific antibody needs to be used in quantitative dot blot analysis (qdb), or then biochemical separation needs to be used before the detection step as there are too many analytes in the sample to perform accurate detection and quantification.

Protein Detection with Antibodies (Immunoassays)

Antibodies to particular proteins or to their modified forms have been used in biochemistry and cell biology studies. These are among the most common tools used by molecular biologists today. There are several specific techniques and protocols that use antibodies for protein detection. The enzyme-linked immunosorbent assay (ELISA) has been used for decades to detect and quantitatively measure proteins in samples. The Western blot can be used for detection and quantification of individual proteins, where in an initial step a complex protein mixture is separated using SDS-PAGE and then the protein of interest is identified using an antibody.

Modified proteins can be studied by developing an antibody specific to that modification. For example, there are antibodies that only recognize certain proteins when they are tyrosine-phosphorylated, known as phospho-specific antibodies. Also, there are antibodies specific to other modifications. These can be used to determine the set of proteins that have undergone the modification of interest.

Antibody-free Protein Detection

While protein detection with antibodies are still very common in molecular biology, also other methods have been developed that do not rely on an antibody. These methods offer various advantages, for instance they are often able to determine the sequence of a protein or peptide, they may have higher throughput than antibody-based and they sometimes can identify and quantify proteins for which no antibody exists.

Detection Methods

One of the earliest method for protein analysis has been Edman degradation (introduced in 1967) where a single peptide is subjected to multiple steps of chemical degradation to resolve its sequence. These methods have mostly been supplanted by technologies that offer higher throughput.

More recent methods use mass spectrometry-based techniques, a development that was made possible by the discovery of "soft ionization" methods such as matrix-assisted laser desorption/ionization (MALDI) and electrospray ionization (ESI) developed in the 1980s. These methods gave rise to the top-down and the bottom-up proteomics workflows where often additional separation is performed before analysis.

Separation Methods

For the analysis of complex biological samples, a reduction of sample complexity is required. This can be performed off-line by one-dimensional or two dimensional separation. More recently, on-line methods have been developed where individual peptides (in bottom-up proteomics approaches) are separated using Reversed-phase chromatography and then directly ionized using ESI; the direct coupling of separation and analysis explains the term "on-line" analysis.

Hybrid Technologies

There are several hybrid technologies that use antibody-based purification of individual analytes and then perform mass spectrometric analysis for identification and quantification. Examples of these methods are the MSIA (mass spectrometric immunoassay) developed by Randall Nelson in 1995 and the SISCAPA (Stable Isotope Standard Capture with Anti-Peptide Antibodies) method, introduced by Leigh Anderson in 2004.

Current Research Methodologies

Fluorescence two-dimensional differential gel electrophoresis (2-D DIGE) can be used to quantify variation in the 2-D DIGE process and establish statistically valid thresholds for assigning quantitative changes between samples.

Comparative proteomic analysis can reveal the role of proteins in complex biological systems, including reproduction. For example, treatment with the insecticide triazophos causes an increase in the content of brown planthopper (*Nilaparvata lugens* (Stål)) male accessory gland proteins (Acps) that can be transferred to females via mating, causing an increase in fecundity (i.e. birth rate) of females. To identify changes in the types of accessory gland proteins (Acps) and reproductive proteins that mated female planthoppers received from male planthoppers, researchers conducted a comparative proteomic analysis of mated *N. lugens* females. The results indicated that these proteins participate in the reproductive process of *N. lugens* adult females and males.

Proteome analysis of *Arabidopsis peroxisomes* has been established as the major unbiased approach for identifying new peroxisomal proteins on a large scale.

There are many approaches to characterizing the human proteome, which is estimated to contain between 20,000 and 25,000 non-redundant proteins. The number of unique protein species will likely increase by between 50,000 and 500,000 due to RNA splicing and proteolysis events, and when post-translational modification are also considered, the total number of unique human proteins is estimated to range in the low millions.

In addition, the first promising attempts to decipher the proteome of animal tumors have recently been reported. This method used as a functional method in *Macrobrachium rosenbergii* protein profiling.

High-throughput Proteomic Technologies

Proteomics has steadily gained momentum over the past decade with the evolution of several approaches. Few of these are new and others build on traditional methods. Mass spectrometry-based methods and micro arrays are the most common technologies for large-scale study of proteins.

Mass Spectrometry and Protein Profiling

There are two mass spectrometry-based methods currently used for protein profiling. The more established and widespread method uses high resolution, two-dimensional electrophoresis to separate proteins from different samples in parallel, followed by selection and staining of differentially expressed proteins to be identified by mass spectrometry. Despite the advances in 2DE and its maturity, it has its limits as well. The central concern is the inability to resolve all the proteins within a sample, given their dramatic range in expression level and differing properties.

The second quantitative approach uses stable isotope tags to differentially label proteins from two different complex mixtures. Here, the proteins within a complex mixture are labeled first isotopically, and then digested to yield labeled peptides. The labeled mixtures are then combined, the peptides separated by multidimensional liquid chromatography and analyzed by tandem mass spectrometry. Isotope coded affinity tag (ICAT) reagents are the widely used isotope tags. In this method, the cysteine residues of proteins get covalently attached to the ICAT reagent, thereby reducing the complexity of the mixtures omitting the non-cysteine residues.

Quantitative proteomics using stable isotopic tagging is an increasingly useful tool in modern development. Firstly, chemical reactions have been used to introduce tags into specific sites or proteins for the purpose of probing specific protein functionalities. The isolation of phosphorylated peptides has been achieved using isotopic labeling and selective chemistries to capture the fraction of protein among the complex mixture. Secondly, the ICAT technology was used to differentiate between partially purified or purified macromolecular complexes such as large RNA polymerase II pre-initiation complex and the proteins complexed with yeast transcription factor. Thirdly, ICAT labeling was recently combined with chromatin isolation to identify and quantify chromatin-associated proteins. Finally ICAT reagents are useful for proteomic profiling of cellular organelles and specific cellular fractions.

Another quantitative approach is the Accurate Mass and Time (AMT) tag approach developed by Richard D. Smith and coworkers at Pacific Northwest National Laboratory. In this approach, increased throughput and sensitivity is achieved by avoiding the need for tandem mass spectrometry, and making use of precisely determined separation time information and highly accurate mass determinations for peptide and protein identifications.

Protein Chips

Balancing the use of mass spectrometers in proteomics and in medicine is the use of protein micro arrays. The aim behind protein micro arrays is to print thousands of protein detecting features for the interrogation of biological samples. Antibody arrays are an example in which a host of different antibodies are arrayed to detect their respective antigens from a sample of human blood. Another approach is the arraying of multiple protein types for the study of properties like protein-DNA, protein-protein and protein-ligand interactions. Ideally, the functional proteomic arrays would contain the entire complement of the proteins of a given organism. The first version of such arrays consisted of 5000 purified proteins from yeast deposited onto glass microscopic slides. Despite the success of first chip, it was a greater challenge for protein arrays to be implemented. Proteins are inherently much more difficult to work with than DNA. They have a broad dynamic range, are less stable than DNA and their structure is difficult to preserve on glass slides, though they are essential for most assays. The global ICAT technology has striking advantages over protein chip technologies.

Reverse-phased Protein Microarrays

This is a promising and newer microarray application for the diagnosis, study and treatment of complex diseases such as cancer. The technology merges laser capture microdissection (LCM) with micro array technology, to produce reverse phase protein microarrays. In this type of microarrays, the whole collection of protein themselves are immobilized with the intent of capturing various stages of disease within an individual patient. When used with LCM, reverse phase arrays can monitor the fluctuating state of proteome among different cell population within a small area of human tissue. This is useful for profiling the status of cellular signaling molecules, among a cross section of tissue that includes both normal and cancerous cells. This approach is useful in monitoring the status of key factors in normal prostate epithelium and invasive prostate cancer tissues. LCM then dissects these tissue and protein lysates were arrayed onto nitrocellulose slides, which were probed with specific antibodies. This method can track all kinds of molecular events and can compare diseased and healthy tissues within the same patient enabling the development of treatment strategies and diagnosis. The ability to acquire proteomics snapshots of neighboring cell populations, using reverse phase microarrays in conjunction with LCM has a number of applications beyond the study of tumors. The approach can provide insights into normal physiology and pathology of all the tissues and is invaluable for characterizing developmental processes and anomalies.

Practical Applications of Proteomics

One major development to come from the study of human genes and proteins has been the identification of potential new drugs for the treatment of disease. This relies on genome and proteome information to identify proteins associated with a disease, which computer software can then use as targets for new drugs. For example, if a certain protein is implicated in a disease, its 3D structure provides the information to design drugs to interfere with the action of the protein. A molecule that fits the active site of an enzyme, but cannot be released by the enzyme, inactivates the enzyme. This is the basis of new drug-discovery tools, which aim to find new drugs to inactivate proteins involved in disease. As genetic differences among individuals are found, researchers expect to use these techniques to develop personalized drugs that are more effective for the individual.

Proteomics is also used to reveal complex plant-insect interactions that help identify candidate genes involved in the defensive response of plants to herbivory.

Interaction Proteomics and Protein Networks

Interaction proteomics is the analysis of protein interactions from scales of binary interactions to proteome- or network-wide. Most proteins function via protein-protein interactions, and one goal of interaction proteomics is to identify binary protein interactions, protein complexes, and interactomes.

Several methods are available to probe protein–protein interactions. While the most traditional method is yeast two-hybrid analysis, a powerful emerging method is affinity purification followed by protein mass spectrometry using tagged protein baits. Other methods include surface plasmon resonance (SPR), protein microarrays, dual polarisation interferometry, microscale thermophoresis and experimental methods such as phage display and *in silico* computational methods.

Knowledge of protein-protein interactions is especially useful in regard to biological networks and systems biology, for example in cell signaling cascades and gene regulatory networks (GRNs, where knowledge of protein-DNA interactions is also informative). Proteome-wide analysis of protein interactions, and integration of these interaction patterns into larger biological networks, is crucial towards understanding systems-level biology.

Expression Proteomics

Expression proteomics includes the analysis of protein expression at larger scale. It helps identify main proteins in a particular sample, and those proteins differentially expressed in related samples—such as diseased vs. healthy tissue. If a protein is found only in a diseased sample then it can be a useful drug target or diagnostic marker. Proteins with same or similar expression profiles may also be functionally related. There are technologies such as 2D-PAGE and mass spectrometry that are used in expression proteomics.

Biomarkers

The National Institutes of Health has defined a biomarker as "a characteristic that is objectively measured and evaluated as an indicator of normal biological processes, pathogenic processes, or pharmacologic responses to a therapeutic intervention."

Understanding the proteome, the structure and function of each protein and the complexities of protein–protein interactions is critical for developing the most effective diagnostic techniques and disease treatments in the future. For example, proteomics is highly useful in identification of candidate biomarkers (proteins in body fluids that are of value for diagnosis), identification of the bacterial antigens that are targeted by the immune response, and identification of possible immunohistochemistry markers of infectious or neoplastic diseases.

An interesting use of proteomics is using specific protein biomarkers to diagnose disease. A number of techniques allow to test for proteins produced during a particular disease, which helps to diagnose the disease quickly. Techniques include western blot, immunohistochemical staining, enzyme linked immunosorbent assay (ELISA) or mass spectrometry. Secretomics, a subfield of proteomics that studies secreted proteins and secretion pathways using proteomic approaches, has recently emerged as an important tool for the discovery of biomarkers of disease.

Proteogenomics

In proteogenomics, proteomic technologies such as mass spectrometry are used for improving gene annotations. Parallel analysis of the genome and the proteome facilitates discovery of post-translational modifications and proteolytic events, especially when comparing multiple species (comparative proteogenomics).

Structural Proteomics

Structural proteomics includes the analysis of protein structures at large-scale. It compares protein structures and helps identify functions of newly discovered genes. The structural analysis also helps to understand that where drugs bind to proteins and also show where proteins interact with each other. This understanding is achieved using different technologies such as X-ray crystallography and NMR spectroscopy.

Bioinformatics for Proteomics (Proteome Informatics)

Much proteomics data is collected with the help of high throughput technologies such as mass spectrometry and microarray. It would often take weeks or months to analyze the data and perform comparisons by hand. For this reason, biologists and chemists are collaborating with computer scientists and mathematicians to create programs and pipeline to computationally analyze the protein data. Using bioinformatics techniques, researchers are capable of faster analysis and data storage. A good place to find lists of current programs and databases is on the ExPASy bioinformatics resource portal. The applications of bioinformatics-based proteomics includes medicine, disease diagnosis, biomarker identification, and many more.

Protein Identification

Mass spectrometry and microarray produce peptide fragmentation information but do not give identification of specific proteins present in the original sample. Due to the lack of specific protein identification, past researchers were forced to decipher the peptide fragments themselves. However, there are currently programs available for protein identification. These programs take the peptide sequences output from mass spectrometry and microarray and return information about matching or similar proteins. This is done through algorithms implemented by the program which perform alignments with proteins from known databases such as UniProt and PROSITE to predict what proteins are in the sample with a degree of certainty.

Protein Structure

The biomolecular structure forms the 3D configuration of the protein. Understanding the protein's structure aids in identification of the protein's interactions and function. It used to be that the 3D structure of proteins could only be determined using X-ray crystallography and NMR spectroscopy. As of 2017, Cryo-electron microscopy is a leading technique, solving difficulties with crystallization (in X-ray crystallography) and conformational ambiguity (in NMR); resolution was 2.2Å as of 2015. Now, through bioinformatics, there are computer programs that can in some cases predict and model the structure of proteins. These programs use the chemical properties of amino acids and structural properties of known proteins to predict the 3D model of sample proteins. This also allows scientists to model protein interactions on a larger scale. In addition, biomedical engineers are developing methods to factor in the flexibility of protein structures to make comparisons and predictions.

Post-translational Modifications

Unfortunately, most programs available for protein analysis are not written for proteins that have undergone post-translational modifications. Some programs will accept post-translational modifications to aid in protein identification but then ignore the modification during further protein analysis. It is important to account for these modifications since they can affect the protein's structure. In turn, computational analysis of post-translational modifications has gained the attention of the scientific community. The current post-translational modification programs are only predictive. Chemists, biologists and computer scientists are working together to create and introduce new pipelines that allow for analysis of post-translational modifications that have been experimentally identified for their effect on the protein's structure and function.

Computational Methods in Studying Protein Biomarkers

One example of the use of bioinformatics and the use of computational methods is the study of protein biomarkers. Computational predictive models have shown that extensive and diverse fe-to-maternal protein trafficking occurs during pregnancy and can be readily detected non-invasively in maternal whole blood. This computational approach circumvented a major limitation, the abundance of maternal proteins interfering with the detection of fetal proteins, to fetal proteomic analysis of maternal blood. Computational models can use fetal gene transcripts previously identified in maternal whole blood to create a comprehensive proteomic network of the term neonate. Such work shows that the fetal proteins detected in pregnant woman's blood originate from a diverse group of tissues and organs from the developing fetus. The proteomic networks contain many biomarkers that are proxies for development and illustrate the potential clinical application of this technology as a way to monitor normal and abnormal fetal development.

An information theoretic framework has also been introduced for biomarker discovery, integrating biofluid and tissue information. This new approach takes advantage of functional synergy between certain biofluids and tissues with the potential for clinically significant findings not possible if tissues and biofluids were considered individually. By conceptualizing tissue-biofluid as information channels, significant biofluid proxies can be identified and then used for guided development of clinical diagnostics. Candidate biomarkers are then predicted based on information transfer criteria across the tissue-biofluid channels. Significant biofluid-tissue relationships can be used to prioritize clinical validation of biomarkers.

Emerging Trends in Proteomics

A number of emerging concepts have the potential to improve current features of proteomics. Obtaining absolute quantification of proteins and monitoring post-translational modifications are the two tasks that impact the understanding of protein function in healthy and diseased cells. For many cellular events, the protein concentrations do not change; rather, their function is modulated by post-translational modifications (PTM). Methods of monitoring PTM are an underdeveloped area in proteomics. Selecting a particular subset of protein for analysis substantially reduces protein complexity, making it advantageous for diagnostic purposes where blood is the starting material. Another important aspect of proteomics, yet not addressed, is that proteomics methods should focus on studying proteins in the context of the environment. The increasing use of chemical cross linkers, introduced into living cells to fix protein-protein, protein-DNA and other interactions, may ameliorate this problem partially. The challenge is to identify suitable methods of preserving relevant interactions. Another goal for studying protein is to develop more sophisticated methods to image proteins and other molecules in living cells and real time.

Proteomics for Systems Biology

Advances in quantitative proteomics would clearly enable more in-depth analysis of cellular systems. Biological systems are subject to a variety of perturbations (cell cycle, cellular differentiation, carcinogenesis, environment (biophysical), etc.). Transcriptional and translational responses to these perturbations results in functional changes to the proteome implicated in response to the stimulus. Therefore, describing and quantifying proteome-wide changes in protein abundance is crucial towards understanding biological phenomenon more holistically, on the level of the entire system. In this way, proteomics can be seen as complementary to genomics, transcriptomics, epigenomics, metabolomics, and other -omics approaches in integrative analyses attempting to define biological phenotypes more comprehensively. As an example, *The Cancer Proteome Atlas* provides quantitative protein expression data for ~200 proteins in over 4,000 tumor samples with matched transcriptomic and genomic data from The Cancer Genome Atlas. Similar datasets in other cell types, tissue types, and species, particularly using deep shotgun mass spectrometry, will be an immensely important resource for research in fields like cancer biology, developmental and stem cell biology, medicine, and evolutionary biology.

Human Plasma Proteome

Characterizing the human plasma proteome has become a major goal in the proteomics arena, but it is also the most challenging proteomes of all human tissues. It contains immunoglobulin, cytokines, protein hormones, and secreted proteins indicative of infection on top of resident, hemostatic proteins. It also contains tissue leakage proteins due to the blood circulation through different tissues in the body. The blood thus contains information on the physiological state of all tissues and, combined with its accessibility, makes the blood proteome invaluable for medical purposes. The blood plasma proteome is thought to beCharacterizing the proteome of blood plasma is a daunting challenge.

The depth of the plasma proteome encompassing a dynamic range of more than 10^{10} between the highest abundant protein (albumin) and the lowest (some cytokines) and is thought to be one of the main challenges for proteomics. Temporal and spatial dynamics further complicate the study of human plasma proteome. The turnover of some proteins is quite faster than others and the protein content of an artery may substantially vary from that of a vein. All these differences make even the simplest proteomic task of cataloging the proteome seem out of reach. To tackle this problem, priorities need to be established. Capturing the most meaningful subset of proteins among the entire proteome to generate a diagnostic tool is one such priority. Secondly, since cancer is associated with enhanced glycosylation of proteins, methods that focus on this part of proteins will also be useful. Again: multiparameter analysis best reveals a pathological state. As these technologies improve, the disease profiles should be continually related to respective gene expression changes. Due to the above-mentioned problems plasma proteomics remained challenging. However, technological advancements and continuous developments seem to result in a revival of plasma proteomics as it was shown recently by a technology called plasma proteome profiling. Due to such technologies researchers were able to investigate inflammation processes in mice, the heretability of plasma proteomes as well as to show the effect of such a common life style change like weight loss on the plasma proteome.

Directed Evolution

Directed evolution (DE, "gelenkte Evolution") is a method used in protein engineering that mimics the process of natural selection to evolve proteins or nucleic acids toward a user-defined goal. It consists of subjecting a gene to iterative rounds of mutagenesis (creating a library of variants), selection (expressing the variants and isolating members with the desired function), and amplification (generating a template for the next round). It can be performed *in vivo* (in living cells), or *in vitro* (free in solution or microdroplet). Directed evolution is used both for protein engineering as an alternative to rationally designing modified proteins, as well as studies of fundamental evolutionary principles in a controlled, laboratory environment.

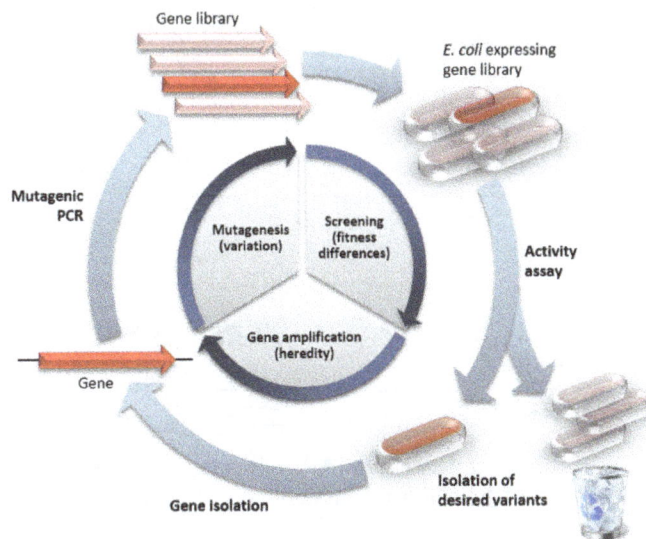

An example of directed evolution with comparison to natural evolution. The inner cycle indicates the 3 stages of the directed evolution cycle with the natural process being mimicked in brackets. The outer circle demonstrates steps a typical experiment. The red symbols indicate functional variants, the pale symbols indicate variants with reduced function.

Principles

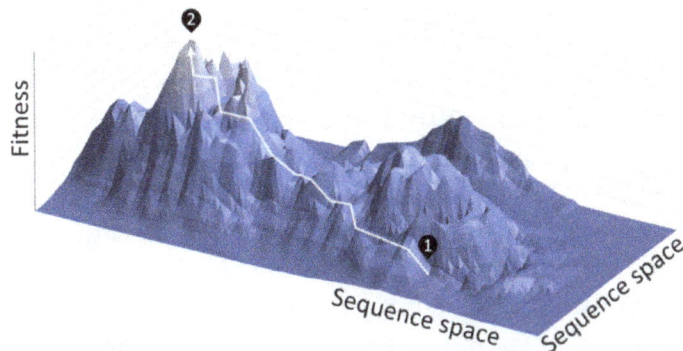

Directed evolution is analogous to climbing a hill on a 'fitness landscape' where elevation represents the desired property. Each round of selection samples mutants on all sides of the starting template (1) and selects the mutant with the highest elevation, thereby climbing the hill. This is repeated until a local summit is reached (2).

Directed evolution is a mimic of the natural evolution cycle in a laboratory setting. Evolution requires three things to occur: variation between replicators, that the variation causes fitness differences upon which selection acts, and that this variation is heritable. In DE, a single gene is evolved by iterative rounds of mutagenesis, selection or screening, and amplification. Rounds of these steps are typically repeated, using the best variant from one round as the template for the next to achieve stepwise improvements.

The likelihood of success in a directed evolution experiment is directly related to the total library size, as evaluating more mutants increases the chances of finding one with the desired properties.

Generating Variation

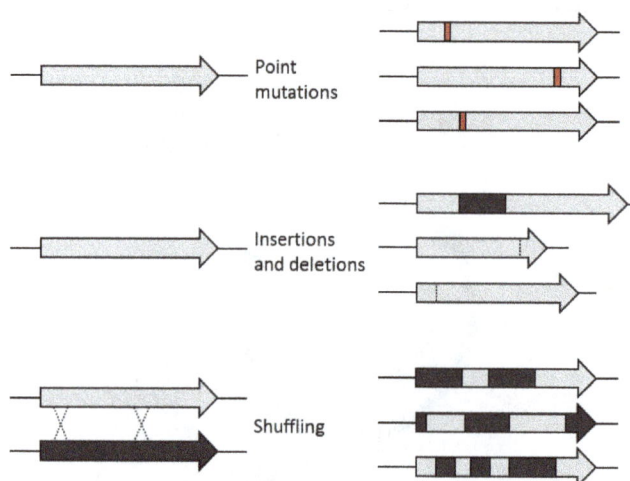

Starting gene (left) and library of variants (right). Point mutations change single nucleotides. Insertions and deletions add or remove sections of DNA. Shuffling recombines segments of two (or more) similar genes.

The first step in performing a cycle of directed evolution is the generation of a library of variant genes. The sequence space for random sequence is vast (10^{130} possible sequences for a 100 amino acid protein) and extremely sparsely populated by functional proteins. Neither experimental, nor natural evolution can ever get close to sampling so many sequences. Of course, natural evolution samples variant sequences close to functional protein sequences and this is imitated in DE by mutagenising an already functional gene. Some calculations suggest it is entirely feasible that for all practical (i.e. functional and structural) purposes, protein sequence space has been fully explored during the course of evolution of life on Earth.

The starting gene can be mutagenised by random point mutations (by chemical mutagens or error prone PCR) and insertions and deletions (by transposons). Gene recombination can be mimicked by DNA shuffling of several sequences (usually of more than 70% homology) to jump into regions of sequence space between the shuffled parent genes. Finally, specific regions of a gene can be systematically randomised for a more focused approach based on structure and function knowledge. Depending on the method, the library generated will vary in the proportion of functional variants it contains. Even if an organism is used to express the gene of interest, by mutagenising only that gene, the rest of the organism's genome remains the same and can be ignored for the evolution experiment (to the extent of providing a constant genetic environment).

Detecting Fitness Differences

The majority of mutations are deleterious and so libraries of mutants tend to mostly have variants with reduced activity. Therefore, a high-throughput assay is vital for measuring activity to find the rare variants with beneficial mutations that improve the desired properties. Two main categories of method exist for isolating functional variants. Selection systems directly couple protein function to survival of the gene, whereas screening systems individually assay each variant and allow a quantitative threshold to be set for sorting a variant or population of variants of a desired activity. Both selection and screening can be performed in living cells (*in vivo* evolution) or performed directly on the protein or RNA without any cells (*in vitro* evolution).

During *in vivo* evolution, each cell (usually bacteria or yeast) is transformed with a plasmid containing a different member of the variant library. In this way, only the gene of interest differs between the cells, with all other genes being kept the same. The cells express the protein either in their cytoplasm or surface where its function can be tested. This format has the advantage of selecting for properties in a cellular environment, which is useful when the evolved protein or RNA is to be used in living organisms. When performed without cells, DE involves using *in vitro* transcription translation to produce proteins or RNA free in solution or compartmentalised in artificial microdroplets. This method has the benefits of being more versatile in the selection conditions (e.g. temperature, solvent), and can express proteins that would be toxic to cells. Furthermore, *in vitro* evolution experiments can generate far larger libraries (up to 10^{15}) because the library DNA need not be inserted into cells (often a limiting step).

Selection

Selection for binding activity is conceptually simple. The target molecule is immobilised on a solid support, a library of variant proteins is flowed over it, poor binders are washed away, and the remaining bound variants recovered to isolate their genes. Binding of an enzyme to immobilised covalent inhibitor has been also used as an attempt to isolate active catalysts. This approach, however, only selects for single catalytic turnover and is not a good model of substrate binding or true substrate reactivity. If an enzyme activity can be made necessary for cell survival, either by synthesizing a vital metabolite, or destroying a toxin, then cell survival is a function of enzyme activity. Such systems are generally only limited in throughput by the transformation efficiency of cells. They are also less expensive and labour-intensive than screening, however they are typically difficult to engineer, prone to artefacts and give no information on the range of activities present in the library.

Screening

An alternative to selection is a screening system. Each variant gene is individually expressed and assayed to quantitatively measure the activity (most often by a colourgenic or fluorogenic product). The variants are then ranked and the experimenter decides which variants to use as templates for the next round of DE. Even the most high throughput assays usually have lower coverage than selection methods but give the advantage of producing detailed information on each one of the screened variants. This disaggregated data can also be used to characterise the distribution of activities in libraries which is not possible in simple selection systems. Screening systems, therefore, have advantages when it comes to experimentally characterising adaptive evolution and fitness landscapes.

Ensuring Heredity

When functional proteins have been isolated, it is necessary that their genes are too, therefore a genotype-phenotype link is required. This can be covalent, such as mRNA display where the mRNA gene is linked to the protein at the end of translation by puromycin. Alternatively the protein and its gene can be co-localised by compartmentalisation in living cells or emulsion droplets. The gene sequences isolated are then amplified by PCR or by transformed host bacteria. Either the single best sequence, or a pool of sequences can be used as the template for the next round of mutagenesis. The repeated cycles of Diversification-Selection-Amplification generate protein variants adapted to the applied selection pressures.

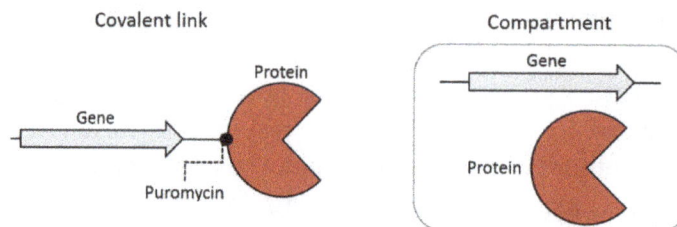

An expressed protein can either be covalently linked to its gene (as in mRNA, left) or compartmentalized with it (cells or artificial compartments, right). Either way ensures that the gene can be isolated based on the activity of the encoded protein.

Comparison to Rational Protein Design

Advantages of Directed Evolution

Rational design of a protein relies on an in-depth knowledge of the protein structure, as well as its catalytic mechanism. Specific changes are then made by site-directed mutagenesis in an attempt to change the function of the protein. A drawback of this is that even when the structure and mechanism of action of the protein are well known, the change due to mutation is still difficult to predict. Therefore, an advantage of DE is that there is no need to understand the mechanism of the desired activity or how mutations would affect it.

Limitations of Directed Evolution

A restriction of directed evolution is that a high-throughput assay is required in order to measure the effects of a large number of different random mutations. This can require extensive research and development before it can be used for directed evolution. Additionally, such assays are often highly specific to monitoring a particular activity and so are not transferable to new DE experiments.

Additionally, selecting for improvement in the assayed function simply generates improvements in the assayed function. To understand how these improvements are achieved, the properties of the evolving enzyme have to be measured. Improvement of the assayed activity can be due to improvements in enzyme catalytic activity or enzyme concentration. There is also no guarantee that improvement on one substrate will improve activity on another. This is particularly important when the desired activity cannot be directly screened or selected for and so a 'proxy' substrate is used. DE can lead to evolutionary specialisation to the proxy without improving the desired activity. Consequently, choosing appropriate screening or selection conditions is vital for successful DE.

Combinatorial Approaches

Combined, 'semi-rational' approaches are being investigated to address the limitations of both rational design and directed evolution. Beneficial mutations are rare, so large numbers of random mutants have to be screened to find improved variants. 'Focussed libraries' concentrate on randomising regions thought to be richer in beneficial mutations for the mutagenesis step of DE. A focussed library contains fewer variants than a traditional random mutagenesis library and so does not require such high-throughput screening.

Creating a focussed library requires some knowledge of which residues in the structure to mutate.

For example, knowledge of the active site of an enzyme may allow just the residues known to interact with the substrate to be randomised. Alternatively, knowledge of which protein regions are variable in nature can guide mutagenesis in just those regions.

Uses

Directed evolution is frequently used for protein engineering as an alternative to rational design, but can also be used to investigate fundamental questions of enzyme evolution.

Protein Engineering

As a protein engineering tool, DE has been most successful in three areas:

1. Improving protein stability for biotechnological use at high temperatures or in harsh solvents.

2. Improving binding affinity of therapeutic antibodies (Affinity maturation) and the activity of de novo designed enzymes.

3. Altering substrate specificity of existing enzymes, (often for use in industry).

Evolution Studies

The study of natural evolution is traditionally based on extant organisms and their genes. However, research is fundamentally limited by the lack of fossils (and particularly the lack of ancient DNA sequences) and incomplete knowledge of ancient environmental conditions. Directed evolution investigates evolution in a controlled system of genes for individual enzymes, ribozymes and replicators (similar to experimental evolution of eukaryotes, prokaryotes and viruses).

DE allows control of selection pressure, mutation rate and environment (both the abiotic environment such as temperature, and the biotic environment, such as other genes in the organism). Additionally, there is a complete record of all evolutionary intermediate genes. This allows for detailed measurements of evolutionary processes, for example epistasis, evolvability, adaptive constraint fitness landscapes, and neutral networks.

Metagenomics

Metagenomics is the study of genetic material recovered directly from environmental samples. The broad field may also be referred to as environmental genomics, ecogenomics or community genomics. While traditional microbiology and microbial genome sequencing and genomics rely upon cultivated clonal cultures, early environmental gene sequencing cloned specific genes (often the 16S rRNA gene) to produce a profile of diversity in a natural sample. Such work revealed that the vast majority of microbial biodiversity had been missed by cultivation-based methods. Recent studies use either "shotgun" or PCR directed sequencing to get largely unbiased samples of all genes from all the members of the sampled communities. Because of its ability to reveal the previously hidden diversity of microscopic life, metagenomics offers a powerful lens for viewing the microbial world that has the potential to revolutionize understanding of the entire living world. As the price of DNA sequencing continues to fall, metagenomics now allows microbial ecology to be investigated at a much greater scale and detail than before.

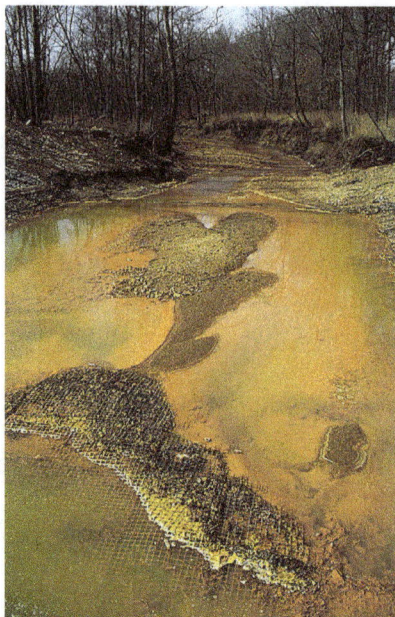

Metagenomics allows the study of microbial communities like those present in this stream receiving acid drainage from surface coal mining.

Etymology

The term "metagenomics" was first used by Jo Handelsman, Jon Clardy, Robert M. Goodman, Sean F. Brady, and others, and first appeared in publication in 1998. The term metagenome referenced the idea that a collection of genes sequenced from the environment could be analyzed in a way analogous to the study of a single genome. Recently, Kevin Chen and Lior Pachter (researchers at the University of California, Berkeley) defined metagenomics as "the application of modern genomics technique without the need for isolation and lab cultivation of individual species".

History

Conventional sequencing begins with a culture of identical cells as a source of DNA. However, early metagenomic studies revealed that there are probably large groups of microorganisms in many environments that cannot be cultured and thus cannot be sequenced. These early studies focused on 16S ribosomal RNA sequences which are relatively short, often conserved within a species, and generally different between species. Many 16S rRNA sequences have been found which do not belong to any known cultured species, indicating that there are numerous non-isolated organisms. These surveys of ribosomal RNA (rRNA) genes taken directly from the environment revealed that cultivation based methods find less than 1% of the bacterial and archaeal species in a sample. Much of the interest in metagenomics comes from these discoveries that showed that the vast majority of microorganisms had previously gone unnoticed.

Early molecular work in the field was conducted by Norman R. Pace and colleagues, who used PCR to explore the diversity of ribosomal RNA sequences. The insights gained from these breakthrough studies led Pace to propose the idea of cloning DNA directly from environmental samples as early as 1985. This led to the first report of isolating and cloning bulk DNA from an environmental sample, published by Pace and colleagues in 1991 while Pace was in the Department of Biology at Indiana University. Considerable efforts ensured that these were not PCR false positives and supported the existence of a complex community of unexplored species. Although this methodology was limited to exploring highly conserved, non-protein coding genes, it did support early microbial morphology-based observations that diversity was far more complex than was known by culturing methods. Soon after that, Healy reported the metagenomic isolation of functional genes from "zoolibraries" constructed from a complex culture of environmental organisms grown in the laboratory on dried grasses in 1995. After leaving the Pace laboratory, Edward DeLong continued in the field and has published work that has largely laid the groundwork for environmental phylogenies based on signature 16S sequences, beginning with his group's construction of libraries from marine samples.

In 2002, Mya Breitbart, Forest Rohwer, and colleagues used environmental shotgun sequencing to show that 200 liters of seawater contains over 5000 different viruses. Subsequent studies showed that there are more than a thousand viral species in human stool and possibly a million different viruses per kilogram of marine sediment, including many bacteriophages. Essentially all of the viruses in these studies were new species. In 2004, Gene Tyson, Jill Banfield, and colleagues at the University of California, Berkeley and the Joint Genome Institute sequenced DNA extracted from an acid mine drainage system. This effort resulted in the complete, or nearly complete, genomes for a handful of bacteria and archaea that had previously resisted attempts to culture them.

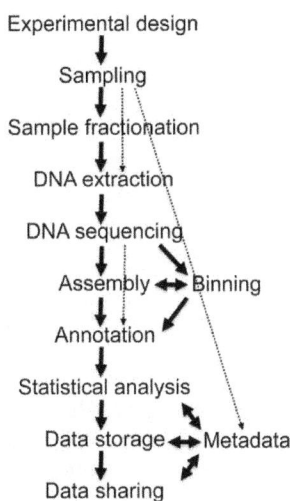

Flow diagram of a typical metagenome project

Beginning in 2003, Craig Venter, leader of the privately funded parallel of the Human Genome Project, has led the Global Ocean Sampling Expedition (GOS), circumnavigating the globe and collecting metagenomic samples throughout the journey. All of these samples are sequenced using shotgun sequencing, in hopes that new genomes (and therefore new organisms) would be identified. The pilot project, conducted in the Sargasso Sea, found DNA from nearly 2000 different species, including 148 types of bacteria never before seen. Venter has circumnavigated the globe and thoroughly explored the West Coast of the United States, and completed a two-year expedition to explore the Baltic, Mediterranean and Black Seas. Analysis of the metagenomic data collected during this journey revealed two groups of organisms, one composed of taxa adapted to environmental conditions of 'feast or famine', and a second composed of relatively fewer but more abundantly and widely distributed taxa primarily composed of plankton.

In 2005 Stephan C. Schuster at Penn State University and colleagues published the first sequences of an environmental sample generated with high-throughput sequencing, in this case massively parallel pyrosequencing developed by 454 Life Sciences. Another early paper in this area appeared in 2006 by Robert Edwards, Forest Rohwer, and colleagues at San Diego State University.

Sequencing

Recovery of DNA sequences longer than a few thousand base pairs from environmental samples was very difficult until recent advances in molecular biological techniques allowed the construction of libraries in bacterial artificial chromosomes (BACs), which provided better vectors for molecular cloning.

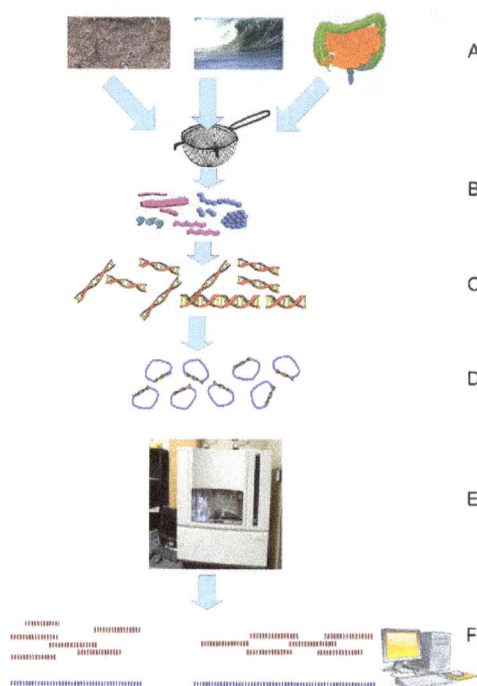

Environmental Shotgun Sequencing (ESS). (A) Sampling from habitat; (B) filtering particles, typically by size; (C) Lysis and DNA extraction; (D) cloning and library construction; (E) sequencing the clones; (F) sequence assembly into contigs and scaffolds.

Shotgun Metagenomics

Advances in bioinformatics, refinements of DNA amplification, and the proliferation of computational power have greatly aided the analysis of DNA sequences recovered from environmental samples, allowing the adaptation of shotgun sequencing to metagenomic samples (known also as whole metagenome shotgun or WMGS sequencing). The approach, used to sequence many cultured microorganisms and the human genome, randomly shears DNA, sequences many short sequences, and reconstructs them into a consensus sequence. Shotgun sequencing reveals genes present in environmental samples. Historically, clone libraries were used to facilitate this sequencing. However, with advances in high throughput sequencing technologies, the cloning step is no longer necessary and greater yields of sequencing data can be obtained without this labour-intensive bottleneck step. Shotgun metagenomics provides information both about which organisms are present and what metabolic processes are possible in the community. Because the collection of DNA from an environment is largely uncontrolled, the most abundant organisms in an environmental sample are most highly represented in the resulting sequence data. To achieve the high coverage needed to fully resolve the genomes of under-represented community members, large samples, often prohibitively so, are needed. On the other hand, the random nature of shotgun sequencing ensures that many of these organisms, which would otherwise go unnoticed using traditional culturing techniques, will be represented by at least some small sequence segments.

High-throughput Sequencing

The first metagenomic studies conducted using high-throughput sequencing used massively parallel 454 pyrosequencing. Three other technologies commonly applied to environmental sampling are the Ion Torrent Personal Genome Machine, the Illumina MiSeq or HiSeq and the Applied Biosystems SOLiD system. These techniques for sequencing DNA generate shorter fragments than Sanger sequencing; Ion Torrent PGM System and 454 pyrosequencing typically produces ~400 bp reads, Illumina MiSeq produces 400-700bp reads (depending on whether paired end options are used), and SOLiD produce 25-75 bp reads. Historically, these read lengths were significantly shorter than the typical Sanger sequencing read length of ~750 bp, however the Illumina technology is quickly coming close to this benchmark. However, this limitation is compensated for by the much larger number of sequence reads. In 2009, pyrosequenced metagenomes generate 200–500 megabases, and Illumina platforms generate around 20–50 gigabases, but these outputs have increased by orders of magnitude in recent years. An additional advantage to high throughput sequencing is that this technique does not require cloning the DNA before sequencing, removing one of the main biases and bottlenecks in environmental sampling.

Bioinformatics

The data generated by metagenomics experiments are both enormous and inherently noisy, containing fragmented data representing as many as 10,000 species. The sequencing of the cow rumen metagenome generated 279 gigabases, or 279 billion base pairs of nucleotide sequence data, while the human gut microbiome gene catalog identified 3.3 million genes assembled from 567.7 gigabases of sequence data. Collecting, curating, and extracting useful biological information from datasets of this size represent significant computational challenges for researchers.

Sequence pre-filtering

The first step of metagenomic data analysis requires the execution of certain pre-filtering steps, including the removal of redundant, low-quality sequences and sequences of probable eukaryotic origin (especially in metagenomes of human origin). The methods available for the removal of contaminating eukaryotic genomic DNA sequences include Eu-Detect and DeConseq.

Assembly

DNA sequence data from genomic and metagenomic projects are essentially the same, but genomic sequence data offers higher coverage while metagenomic data is usually highly non-redundant. Furthermore, the increased use of second-generation sequencing technologies with short read lengths means that much of future metagenomic data will be error-prone. Taken in combination, these factors make the assembly of metagenomic sequence reads into genomes difficult and unreliable. Misassemblies are caused by the presence of repetitive DNA sequences that make assembly especially difficult because of the difference in the relative abundance of species present in the sample. Misassemblies can also involve the combination of sequences from more than one species into chimeric contigs.

There are several assembly programs, most of which can use information from paired-end tags in order to improve the accuracy of assemblies. Some programs, such as Phrap or Celera Assembler, were designed to be used to assemble single genomes but nevertheless produce good results when assembling metagenomic data sets. Other programs, such as Velvet assembler, have been optimized for the shorter reads produced by second-generation sequencing through the use of de Bruijn graphs. The use of reference genomes allows researchers to improve the assembly of the most abundant microbial species, but this approach is limited by the small subset of microbial phyla for which sequenced genomes are available. After an assembly is created, an additional challenge is "metagenomic deconvolution", or determining which sequences come from which species in the sample.

Gene Prediction

Metagenomic analysis pipelines use two approaches in the annotation of coding regions in the assembled contigs. The first approach is to identify genes based upon homology with genes that are already publicly available in sequence databases, usually by simple BLAST searches. This type of approach is implemented in the program MEGAN4. The second, *ab initio*, uses intrinsic features of the sequence to predict coding regions based upon gene training sets from related organisms. This is the approach taken by programs such as GeneMark and GLIMMER. The main advantage of *ab initio* prediction is that it enables the detection of coding regions that lack homologs in the sequence databases; however, it is most accurate when there are large regions of contiguous genomic DNA available for comparison.

Species Diversity

Gene annotations provide the "what", while measurements of species diversity provide the "who". In order to connect community composition and function in metagenomes, sequences must be binned. Binning is the process of associating a particular sequence with an organism. In similarity-based binning, methods such as BLAST are used to rapidly search for phylogenetic markers or otherwise similar sequences in existing public databases. This approach is implemented in MEGAN. Another tool, PhymmBL, uses interpolated Markov models to assign reads. MetaPhlAn and AMPHORA are methods based on unique clade-specific markers for estimating organismal relative abundances with improved computational performances. Recent methods, such as SLIMM, use read coverage landscape of individual reference genomes to minimize false-positive hits and get reliable relative abundances. In composition based binning, methods use intrinsic features of the sequence, such as oligonucleotide frequencies or codon usage bias. Once sequences are binned, it is possible to carry out comparative analysis of diversity and richness.

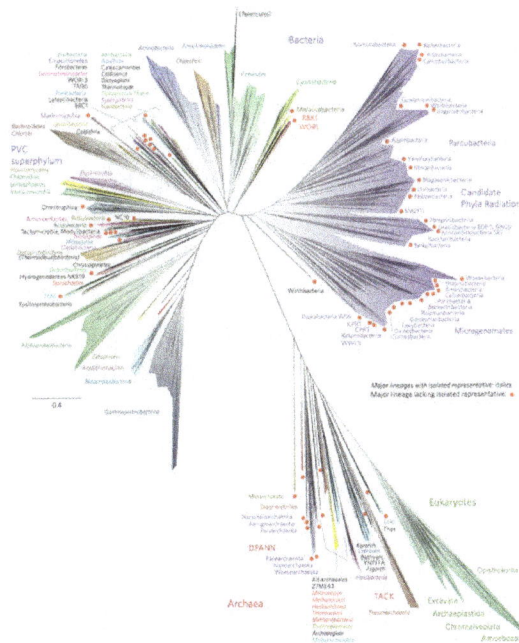

A 2016 representation of the tree of life

Data Integration

The massive amount of exponentially growing sequence data is a daunting challenge that is complicated by the complexity of the metadata associated with metagenomic projects. Metadata includes detailed information about the three-dimensional (including depth, or height) geography and environmental features of the sample, physical data about the sample site, and the methodology of the sampling. This information is necessary both to ensure replicability and to enable downstream analysis. Because of its importance, metadata and collaborative data review and curation require standardized data formats located in specialized databases, such as the Genomes OnLine Database (GOLD).

Several tools have been developed to integrate metadata and sequence data, allowing downstream comparative analyses of different datasets using a number of ecological indices. In 2007, Folker Meyer and Robert Edwards and a team at Argonne National Laboratory and the University of Chicago released the Metagenomics Rapid Annotation using Subsystem Technology server (MG-RAST) a community resource for metagenome data set analysis. As of June 2012 over 14.8 terabases (14×10^{12} bases) of DNA have been analyzed, with more than 10,000 public data sets freely available for comparison within MG-RAST. Over 8,000 users now have submitted a total of 50,000 metagenomes to MG-RAST. The Integrated Microbial Genomes/Metagenomes (IMG/M) system also provides a collection of tools for functional analysis of microbial communities based on their metagenome sequence, based upon reference isolate genomes included from the Integrated Microbial Genomes (IMG) system and the Genomic Encyclopedia of Bacteria and Archaea (GEBA) project.

One of the first standalone tools for analysing high-throughput metagenome shotgun data was MEGAN (MEta Genome ANalyzer). A first version of the program was used in 2005 to analyse the metagenomic context of DNA sequences obtained from a mammoth bone. Based on a BLAST comparison against a reference database, this tool performs both taxonomic and functional binning, by placing the reads onto the nodes of the NCBI taxonomy using a simple lowest common ancestor (LCA) algorithm or onto the nodes of the SEED or KEGG classifications, respectively.

With the advent of fast and inexpensive sequencing instruments, the growth of databases of DNA sequences is now exponential (e.g., the NCBI GenBank database). Faster and efficient tools are needed to keep pace with the high-throughput sequencing, because the BLAST-based approaches such as MG-RAST or MEGAN run slowly to annotate large samples (e.g., several hours to process a small/medium size dataset/sample). Thus, ultra-fast classifiers have recently emerged, thanks to more affordable powerful servers. These tools can perform the taxonomic annotation at extremely high speed, for example CLARK (according to CLARK's authors, it can classify accurately "32 million metagenomic short reads per minute"). At such a speed, a very large dataset/sample of a billion short reads can be processed in about 30 minutes.

Comparative Metagenomics

Comparative analyses between metagenomes can provide additional insight into the function of complex microbial communities and their role in host health. Pairwise or multiple comparisons between metagenomes can be made at the level of sequence composition (comparing GC-content or genome size), taxonomic diversity, or functional complement. Comparisons of population structure and phylogenetic diversity can be made on the basis of 16S and other phylogenetic marker genes, or—in the case of low-diversity communities—by genome reconstruction from the metagenomic dataset. Functional comparisons between metagenomes may be made by comparing sequences against reference databases such as COG or KEGG, and tabulating the abundance by category and evaluating any differences for statistical significance. This gene-centric approach emphasizes the functional complement of the *community* as a whole rather than taxonomic groups, and shows that the functional complements are analogous under similar environmental conditions. Consequently, metadata on the environmental context of the metagenomic sample is especially important in comparative analyses, as it provides researchers with the ability to study the effect of habitat upon community structure and function.

Additionally, several studies have also utilized oligonucleotide usage patterns to identify the differences across diverse microbial communities. Examples of such methodologies include the dinucleotide relative abundance approach by Willner et al. and the HabiSign approach of Ghosh et al. This latter study also indicated that differences in tetranucleotide usage patterns can be used to identify genes (or metagenomic reads) originating from specific habitats. Additionally some methods as TriageTools or Compareads detect similar reads between two read sets. The similarity measure they apply on reads is based on a number of identical words of length *k* shared by pairs of reads.

A key goal in comparative metagenomics is to identify microbial group(s) which are responsible for conferring specific characteristics to a given environment. However, due to issues in the sequencing technologies artifacts need to be accounted for like in metagenomeSeq. Others have characterized inter-microbial interactions between the resident microbial groups. A GUI-based comparative metagenomic analysis application called Community-Analyzer has been developed by Kuntal et al. which implements a correlation-based graph layout algorithm that not only facilitates a quick visualization of the differences in the analyzed microbial communities (in terms of their taxonomic composition), but also provides insights into the inherent inter-microbial interactions occurring therein. Notably, this layout algorithm also enables grouping of the metagenomes based on the probable inter-microbial interaction patterns rather than simply comparing abundance values of various taxonomic groups. In addition, the tool implements several interactive GUI-based functionalities that enable users to perform standard comparative analyses across microbiomes.

Data Analysis

Community Metabolism

In many bacterial communities, natural or engineered (such as bioreactors), there is significant division of labor in metabolism (Syntrophy), during which the waste products of some organisms are metabolites for others. In one such system, the methanogenic bioreactor, functional stability requires the presence of several syntrophic species (Syntrophobacterales and Synergistia) working together in order to turn raw resources into fully metabolized waste (methane). Using comparative gene studies and expression experiments with microarrays or proteomics researchers can piece together a metabolic network that goes beyond species boundaries. Such studies require detailed knowledge about which versions of which proteins are coded by which species and even by which strains of which species. Therefore, community genomic information is another fundamental tool (with metabolomics and proteomics) in the quest to determine how metabolites are transferred and transformed by a community.

Metatranscriptomics

Metagenomics allows researchers to access the functional and metabolic diversity of microbial communities, but it cannot show which of these processes are active. The extraction and analysis of metagenomic mRNA (the metatranscriptome) provides information on the regulation and expression profiles of complex communities. Because of the technical difficulties (the short half-life of mRNA, for example) in the collection of environmental RNA there have been relatively few *in situ* metatranscriptomic studies of microbial communities to date. While originally limited to microarray technology, metatranscriptomcs studies have made use of direct high-throughput cDNA sequencing to provide whole-genome expression and quantification of a microbial community, as first employed by Leininger *et al.* (2006) in their analysis of ammonia oxidation in soils.

Viruses

Metagenomic sequencing is particularly useful in the study of viral communities. As viruses lack a shared universal phylogenetic marker (as 16S RNA for bacteria and archaea, and 18S RNA for eukarya), the only way to access the genetic diversity of the viral community from an environmental sample is through metagenomics. Viral metagenomes (also called viromes) should thus provide more and more information about viral diversity and evolution. For example, a metagenomic pipeline called Giant Virus Finder showed the first evidence of existence of giant viruses in a saline desert and in Antarctic dry valleys.

Applications

Metagenomics has the potential to advance knowledge in a wide variety of fields. It can also be applied to solve practical challenges in medicine, engineering, agriculture, sustainability and ecology.

Infectious Disease Diagnosis

Differentiating between infectious and non-infectious illness, and identifying the underlying etiology of infection, can be quite challenging. For example, more than half of cases of encephalitis remain undiagnosed, despite extensive testing using state-of-the-art clinical laboratory methods. Metagenomic sequencing shows promise as a sensitive and rapid method to diagnose infection by comparing genetic material found in a patient's sample to a database of thousands of bacteria, viruses, and other pathogens.

Gut Microbe Characterization

Microbial communities play a key role in preserving human health, but their composition and the mechanism by which they do so remains mysterious. Metagenomic sequencing is being used to characterize the microbial communities from 15-18 body sites from at least 250 individuals. This is part of the Human Microbiome initiative with primary goals to determine if there is a core human microbiome, to understand the changes in the human microbiome that can be correlated with human health, and to develop new technological and bioinformatics tools to support these goals.

Another medical study as part of the MetaHit (Metagenomics of the Human Intestinal Tract) project consisted of 124 individuals from Denmark and Spain consisting of healthy, overweight, and irritable bowel disease patients. The study attempted to categorize the depth and phylogenetic diversity of gastrointestinal bacteria. Using Illumina GA sequence data and SOAPdenovo, a de Bruijn graph-based tool specifically designed for assembly short reads, they were able to generate 6.58 million contigs greater than 500 bp for a total contig length of 10.3 Gb and a N50 length of 2.2 kb.

The study demonstrated that two bacterial divisions, Bacteroidetes and Firmicutes, constitute over 90% of the known phylogenetic categories that dominate distal gut bacteria. Using the relative gene frequencies found within the gut these researchers identified 1,244 metagenomic clusters that are critically important for the health of the intestinal tract. There are two types of functions in these range clusters: housekeeping and those specific to the intestine. The housekeeping gene clusters are required in all bacteria and are often major players in the main metabolic pathways including central carbon metabolism and amino acid synthesis. The gut-specific functions include adhesion to host proteins and the harvesting of sugars from globoseries glycolipids. Patients with irritable bowel syndrome were shown to exhibit 25% fewer genes and lower bacterial diversity than individuals not suffering from irritable bowel syndrome indicating that changes in patients' gut biome diversity may be associated with this condition.

While these studies highlight some potentially valuable medical applications, only 31-48.8% of the reads could be aligned to 194 public human gut bacterial genomes and 7.6-21.2% to bacterial genomes available in GenBank which indicates that there is still far more research necessary to capture novel bacterial genomes.

Biofuel

Biofuels are fuels derived from biomass conversion, as in the conversion of cellulose contained in corn stalks, switchgrass, and other biomass into cellulosic ethanol. This process is dependent upon microbial consortia that transform the cellulose into sugars, followed by the fermentation of the sugars into ethanol. Microbes also produce a variety of sources of bioenergy including methane and hydrogen.

Bioreactors allow the observation of microbial communities as they convert biomass into cellulosic ethanol.

The efficient industrial-scale deconstruction of biomass requires novel enzymes with higher productivity and lower cost. Metagenomic approaches to the analysis of complex microbial communities allow the targeted screening of enzymes with industrial applications in biofuel production, such as glycoside hydrolases. Furthermore, knowledge of how these microbial communities function is required to control them, and metagenomics is a key tool in their understanding. Metagenomic approaches allow comparative analyses between convergent microbial systems like biogas fermenters or insect herbivores such as the fungus garden of the leafcutter ants.

Environmental Remediation

Metagenomics can improve strategies for monitoring the impact of pollutants on ecosystems and for cleaning up contaminated environments. Increased understanding of how microbial communities cope with pollutants improves assessments of the potential of contaminated sites to recover from pollution and increases the chances of bioaugmentation or biostimulation trials to succeed.

Biotechnology

Microbial communities produce a vast array of biologically active chemicals that are used in competition and communication. Many of the drugs in use today were originally uncovered in microbes; recent progress in mining the rich genetic resource of non-culturable microbes has led to the discovery of new genes, enzymes, and natural products. The application of metagenomics has allowed the development of commodity and fine chemicals, agrochemicals and pharmaceuticals where the benefit of enzyme-catalyzed chiral synthesis is increasingly recognized.

Two types of analysis are used in the bioprospecting of metagenomic data: function-driven screening for an expressed trait, and sequence-driven screening for DNA sequences of interest. Function-driven analysis seeks to identify clones expressing a desired trait or useful activity, followed by biochemical characterization and sequence analysis. This approach is limited by availability of a suitable screen and the requirement that the desired trait be expressed in the host cell. Moreover, the low rate of discovery (less than one per 1,000 clones screened) and its labor-intensive nature further limit this approach. In contrast, sequence-driven analysis uses conserved DNA sequences to design PCR primers to screen clones for the sequence of interest. In comparison to cloning-based approaches, using a sequence-only approach further reduces the amount of bench work required. The application of massively parallel sequencing also greatly increases the amount of sequence data generated, which require high-throughput bioinformatic analysis pipelines. The sequence-driven approach to screening is limited by the breadth and accuracy of gene functions present in public sequence databases. In practice, experiments make use of a combination of both functional and sequence-based approaches based upon the function of interest, the complexity of the sample to be screened, and other factors.

Agriculture

The soils in which plants grow are inhabited by microbial communities, with one gram of soil containing around 10^9-10^{10} microbial cells which comprise about one gigabase of sequence information. The microbial communities which inhabit soils are some of the most complex known to science, and remain poorly understood despite their economic importance. Microbial consortia perform a wide variety of ecosystem services necessary for plant growth, including fixing atmospheric nitrogen, nutrient cycling, disease suppression, and sequester iron and other metals. Functional metagenomics strategies are being used to explore the interactions between plants and microbes through cultivation-independent study of these microbial communities. By allowing insights into the role of previously uncultivated or rare community members in nutrient cycling and the promotion of plant growth, metagenomic approaches can contribute to improved disease detection in crops and livestock and the adaptation of enhanced farming practices which improve crop health by harnessing the relationship between microbes and plants.

Ecology

Metagenomics can provide valuable insights into the functional ecology of environmental communities. Metagenomic analysis of the bacterial consortia found in the defecations of Australian sea lions suggests that nutrient-rich sea lion faeces may be an important nutrient source for coastal ecosystems. This is because the bacteria that are expelled simultaneously with the defecations are adept at breaking down the nutrients in the faeces into a bioavailable form that can be taken up into the food chain.

DNA sequencing can also be used more broadly to identify species present in a body of water, debris filtered from the air, or sample of dirt. This can establish the range of invasive species and endangered species, and track seasonal populations.

References

- Tureček F (2002). "Mass spectrometry in coupling with affinity capture-release and isotope-coded affinity tags for quantitative protein analysis". Journal of Mass Spectrometry. 37 (1): 1–14. PMID 11813306. doi:10.1002/jms.275

- De Mol, NJ (2012). "Surface plasmon resonance for proteomics". Methods in molecular biology (Clifton, N.J.).

Methods in Molecular Biology. 800: 33–53. ISBN 978-1-61779-348-6. PMID 21964781. doi:10.1007/978-1-61779-349-3_4

- Schloss, Patrick D; Jo Handelsman (June 2003). "Biotechnological prospects from metagenomics" (PDF). Current Opinion in Biotechnology. 14 (3): 303–310. ISSN 0958-1669. PMID 12849784. doi:10.1016/S0958-1669(03)00067-3. Retrieved 3 January 2012

- Cox Jü, Mann M (2007). "Is Proteomics the New Genomics?". Cell. 130 (3): 395–8. PMID 17693247. doi:10.1016/j.cell.2007.07.032

- Chen S, Zhang Q, Wu X, Schultz PG, Ding S (2004). "Dedifferentiation of Lineage-Committed Cells by a Small Molecule". Journal of the American Chemical Society. 126 (2): 410–1. PMID 14719906. doi:10.1021/ja037390k

- Marco, D, ed. (2011). Metagenomics: Current Innovations and Future Trends. Caister Academic Press. ISBN 978-1-904455-87-5

- Luheshi LM, Crowther DC, Dobson CM (2008). "Protein misfolding and disease: from the test tube to the organism". Current Opinion in Chemical Biology. 12 (1): 25–31. PMID 18295611. doi:10.1016/j.cbpa.2008.02.011

- Lipovsek, D; Plückthun, A (July 2004). "In-vitro protein evolution by ribosome display and mRNA display.". Journal of immunological methods. 290 (1–2): 51–67. PMID 15261571. doi:10.1016/j.jim.2004.04.008

- Committee on Metagenomics: Challenges and Functional Applications, National Research Council (2007). The New Science of Metagenomics: Revealing the Secrets of Our Microbial Planet. Washington, D.C.: The National Academies Press. ISBN 0-309-10676-1

- "APAF - The Australian Proteome Analysis Facility - APAF - The Australian Proteome Analysis Facility". www.proteome.org.au. Retrieved 2017-02-06

- Kent S (June 2006). "Obituary: Bruce Merrifield (1921–2006)". Nature. 441 (7095): 824. Bibcode:2006Natur.441..824K. PMID 16778881. doi:10.1038/441824a

- Chen, K.; Pachter, L. (2005). "Bioinformatics for Whole-Genome Shotgun Sequencing of Microbial Communities". PLoS Computational Biology. 1 (2): e24. PMC 1185649 . PMID 16110337. doi:10.1371/journal.pcbi.0010024

- Nelson KE and White BA (2010). "Metagenomics and Its Applications to the Study of the Human Microbiome". Metagenomics: Theory, Methods and Applications. Caister Academic Press. ISBN 978-1-904455-54-7

- Tawfik DS, Griffiths AD (1998). "Man-made cell-like compartments for molecular evolution". Nature Biotechnology. 16 (7): 652–6. PMID 9661199. doi:10.1038/nbt0798-652

- Pace, NR; Delong, EF; Pace, NR (1991). "Analysis of a marine picoplankton community by 16S rRNA gene cloning and sequencing". Journal of Bacteriology. 173 (14): 4371–4378. PMC 208098 . PMID 2066334

- Konopka, A. (2009). "What is microbial community ecology?". The ISME Journal. 3 (11): 1223–1230. PMID 19657372. doi:10.1038/ismej.2009.88

Amino Acids in Bioorganic Chemistry

Amino acids are the building block of proteins. The specific pattern of an amino acid depend son the sequence of the protein. Organic molecules which consist of acidic carboxyl group and an organic R group are known as amino acids. The main elements of it are oxygen, hydrogen, nitrogen and carbon. This chapter has been carefully written to provide an easy understanding of the varied facets of amino acids.

Amino Acids

- Amino acids are building blocks of proteins.

- Proteins are composed of 20 different amino acid (encoded by standard genetic code, construct proteins in all species).

- Their molecules containing both amino and carboxyl groups attached to the same a-carbon

- 19 are 1°-amines, 1 (proline) is a 2°-amine

- 19 amino acids are "chiral" and 1 (glycine) is achiral (R=H)

- The configuration of the "natural" amino acids is L (L-a-amino acids).

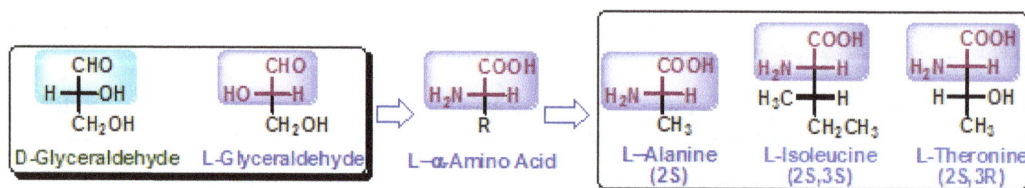

- Their chemical structure influences three dimensional structures of proteins.

- They are important intermediates in metabolism (porphyrins, purines, pyrimidines, creatin, urea etc).

- They can have hormonal and catalytic function.

- Several genetic disorders are cause in amino acid metabolism errors (aminoaciduria - presence of amino acids in urine)

Amino acids are organic compounds containing amine ($-NH_2$) and carboxyl ($-COOH$) functional groups, along with a side chain (R group) specific to each amino acid. The key elements of an amino acid are carbon, hydrogen, oxygen, and nitrogen, although other elements are found in the side chains of certain amino acids. About 500 amino acids are known (though only 20 appear in the genetic code) and can be classified in many ways. They can be classified according to the core structural functional groups' locations as alpha- (α-), beta- (β-), gamma- (γ-) or delta- (δ-) amino acids; other categories relate to polarity, pH level, and side chain group type (aliphatic, acyclic, aromatic, containing hydroxyl or sulfur, etc.). In the form of proteins, amino acid residues form the second-largest component (water is the largest) of human muscles and other tissues. Beyond their role as residues in proteins, amino acids participate in a number of processes such as neurotransmitter transport and biosynthesis.

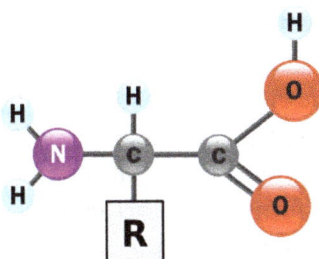

The structure of an alpha amino acid in its un-ionized form

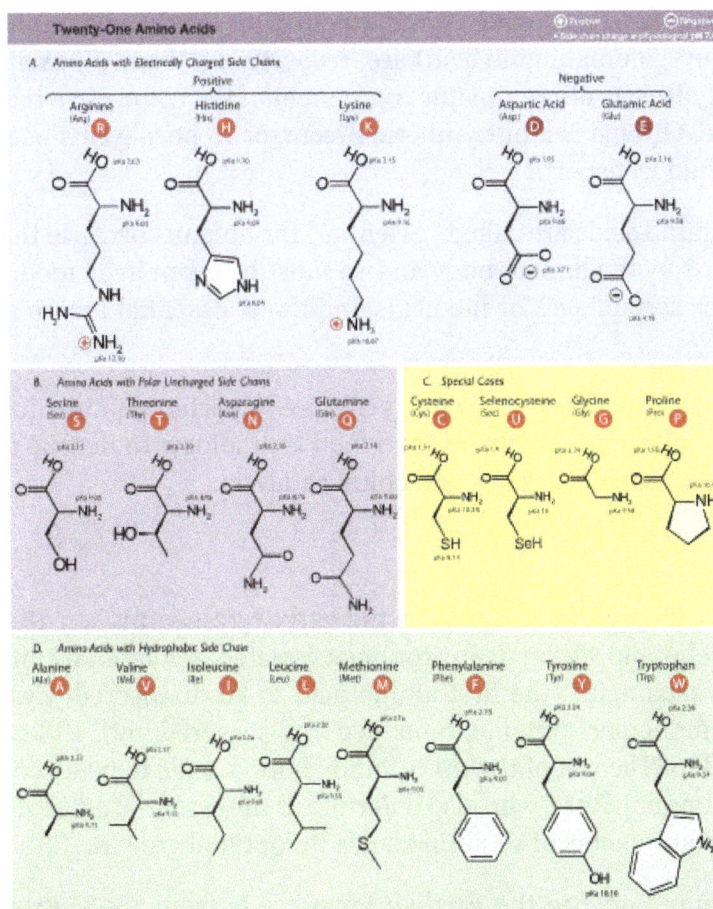

The 21 proteinogenic α-amino acids found in eukaryotes, grouped according to their side chains' pK$_a$ values and charges carried at physiological pH 7.4

In biochemistry, amino acids having both the amine and the carboxylic acid groups attached to the first (alpha-) carbon atom have particular importance. They are known as 2-, alpha-, or α-amino acids (generic formula $H_2NCHRCOOH$ in most cases, where R is an organic substituent known as a "side chain"); often the term "amino acid" is used to refer specifically to these. They include the 22 proteinogenic ("protein-building") amino acids, which combine into peptide chains ("polypeptides") to form the building-blocks of a vast array of proteins. These are all L-stereoisomers ("left-handed" isomers), although a few D-amino acids ("right-handed") occur in bacterial envelopes, as a neuromodulator (D-serine), and in some antibiotics. Twenty of the proteinogenic amino acids are encoded directly by triplet codons in the genetic code and are known as "standard" amino acids. The other two ("non-standard" or "non-canonical") are selenocysteine (present in many noneukaryotes as well as most eukaryotes, but not coded directly by DNA), and pyrrolysine (found only in some archea and one bacterium). Pyrrolysine and selenocysteine are encoded via variant codons; for example, selenocysteine is encoded by stop codon and SECIS element. N-formylmethionine (which is often the initial amino acid of proteins in bacteria, mitochondria, and chloroplasts) is generally considered as a form of methionine rather than as a separate proteinogenic amino acid. Codon–tRNA combinations not found in nature can also be used to "expand" the genetic code and create novel proteins known as alloproteins incorporating non-proteinogenic amino acids.

Many important proteinogenic and non-proteinogenic amino acids have biological functions. For example, in the human brain, glutamate (standard glutamic acid) and gamma-amino-butyric acid ("GABA", non-standard gamma-amino acid) are, respectively, the main excitatory and inhibitory neurotransmitters. Hydroxyproline, a major component of the connective tissue collagen), is synthesised from proline. Glycine is a biosynthetic precursor to porphyrins used in red blood cells. Carnitine is used in lipid transport.

Nine proteinogenic amino acids are called "essential" for humans because they cannot be created from other compounds by the human body and so must be taken in as food. Others may be conditionally essential for certain ages or medical conditions. Essential amino acids may also differ between species.

Because of their biological significance, amino acids are important in nutrition and are commonly used in nutritional supplements, fertilizers, and food technology. Industrial uses include the production of drugs, biodegradable plastics, and chiral catalysts.

History

The first few amino acids were discovered in the early 19th century. In 1806, French chemists Louis-Nicolas Vauquelin and Pierre Jean Robiquet isolated a compound in asparagus that was subsequently named asparagine, the first amino acid to be discovered. Cystine was discovered in 1810, although its monomer, cysteine, remained undiscovered until 1884. Glycine and leucine were discovered in 1820. The last of the 20 common amino acids to be discovered was threonine in 1935 by William Cumming Rose, who also determined the essential amino acids and established the minimum daily requirements of all amino acids for optimal growth.

Usage of the term *amino acid* in the English language is from 1898. Proteins were found to yield amino acids after enzymatic digestion or acid hydrolysis. In 1902, Emil Fischer and Franz Hofmeister proposed that proteins are the result of the formation of bonds between the amino group of one amino acid with the carboxyl group of another, in a linear structure that Fischer termed "peptide".

General Structure

In the structure shown at the top of the page, R represents a side chain specific to each amino acid. The carbon atom next to the carboxyl group (which is therefore numbered 2 in the carbon chain starting from that functional group) is called the α–carbon. Amino acids containing an amino group bonded directly to the alpha carbon are referred to as *alpha amino acids*. These include amino acids such as proline which contain secondary amines, which used to be often referred to as "imino acids".

Isomerism

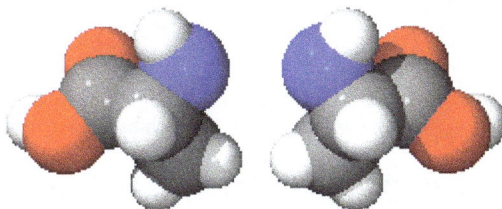

The two enantiomers of alanine, D-alanine and L-alanine

The alpha amino acids are the most common form found in nature, but only when occurring in the L-isomer. The alpha carbon is a chiral carbon atom, with the exception of glycine which has two indistinguishable hydrogen atoms on the alpha carbon. Therefore, all alpha amino acids but glycine can exist in either of two enantiomers, called L or D amino acids, which are mirror images of each other. While L-amino acids represent all of the amino acids found in proteins during translation in the ribosome, D-amino acids are found in some proteins produced by enzyme posttranslational modifications after translation and translocation to the endoplasmic reticulum, as in exotic sea-dwelling organisms such as cone snails. They are also abundant components of the peptidoglycan cell walls of bacteria, and D-serine may act as a neurotransmitter in the brain. D-amino acids are used in racemic crystallography to create centrosymmetric crystals, which (depending on the protein) may allow for easier and more robust protein structure determination. The L and D convention for amino acid configuration refers not to the optical activity of the amino acid itself but rather to the optical activity of the isomer of glyceraldehyde from which that amino acid can, in theory, be synthesized (D-glyceraldehyde is dextrorotatory; L-glyceraldehyde is levorotatory). In alternative fashion, the *(S)* and *(R)* designators are used to indicate the absolute stereochemistry. Almost all of the amino acids in proteins are *(S)* at the α carbon, with cysteine being *(R)* and glycine non-chiral. Cysteine has its side chain in the same geometric position as the other amino acids, but the *R/S* terminology is reversed because of the higher atomic number of sulfur compared to the carboxyl oxygen gives the side chain a higher priority, whereas the atoms in most other side chains give them lower priority.

Side Chains

In amino acids that have a carbon chain attached to the α–carbon (such as lysine) the carbons are labeled in order as α, β, γ, δ, and so on. In some amino acids, the amine group is attached to the β or γ-carbon, and these are therefore referred to as *beta* or *gamma amino acids*.

Amino acids are usually classified by the properties of their side chain into four groups. The side chain can make an amino acid a weak acid or a weak base, and a hydrophile if the side chain is polar or a hydrophobe if it is nonpolar. The chemical structures of the 22 standard amino acids, along with their chemical properties, are described more fully in the article on these proteinogenic amino acids.

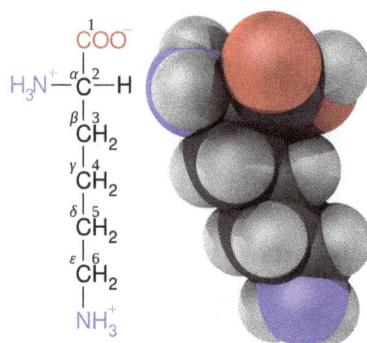

Lysine with carbon atoms labeled

The phrase "branched-chain amino acids" or BCAA refers to the amino acids having aliphatic side chains that are non-linear; these are leucine, isoleucine, and valine. Proline is the only proteinogenic amino acid whose side-group links to the α-amino group and, thus, is also the only proteinogenic amino acid containing a secondary amine at this position. In chemical terms, proline is, therefore, an imino acid, since it lacks a primary amino group, although it is still classed as an amino acid in the current biochemical nomenclature, and may also be called an "N-alkylated alpha-amino acid".

Zwitterions

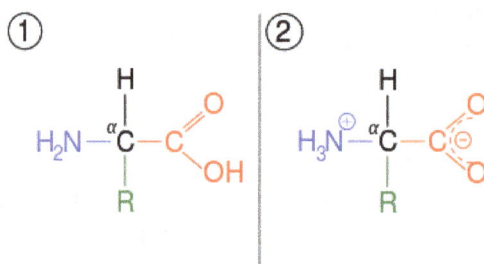

An amino acid in its (1) un-ionized and (2) zwitterionic forms

The α-carboxylic acid group of amino acids is a weak acid, meaning that it releases a hydron (such as a proton) at moderate pH values. In other words, carboxylic acid groups ($-CO_2H$) can be deprotonated to become negative carboxylates ($-CO_2^-$). The negatively charged carboxylate ion predominates at pH values greater than the pKa of the carboxylic acid group (mean for the 20 common amino acids is about 2.2, see the table of amino acid structures above). In a complementary fashion, the α-amine of amino acids is a weak base, meaning that it accepts a proton at moderate pH values. In other words, α-amino groups (NH_2-) can be protonated to become positive α-ammonium groups ($^+NH_3-$). The positively charged α-ammonium group predominates at pH values less than the pKa of the α-ammonium group (mean for the 20 common α-amino acids is about 9.4).

Because all amino acids contain amine and carboxylic acid functional groups, they share amphiprotic properties. Below pH 2.2, the predominant form will have a neutral carboxylic acid group and a positive α-ammonium ion (net charge +1), and above pH 9.4, a negative carboxylate and neutral α-amino group (net charge −1). But at pH between 2.2 and 9.4, an amino acid usually contains both a negative carboxylate and a positive α-ammonium group, as shown in structure (2) on the right, so has net zero charge. This molecular state is known as a zwitterion, from the German Zwitter meaning *hermaphrodite* or *hybrid*. The fully neutral form (structure (1) on the right) is a very minor species in aqueous solution throughout the pH range (less than 1 part in 10⁷). Amino acids exist as zwitterions also in the solid phase, and crystallize with salt-like properties unlike typical organic acids or amines.

Isoelectric Point

The variation in titration curves when the amino acids are grouped by category can be seen here. With the exception of tyrosine, using titration to differentiate between hydrophobic amino acids is problematic.

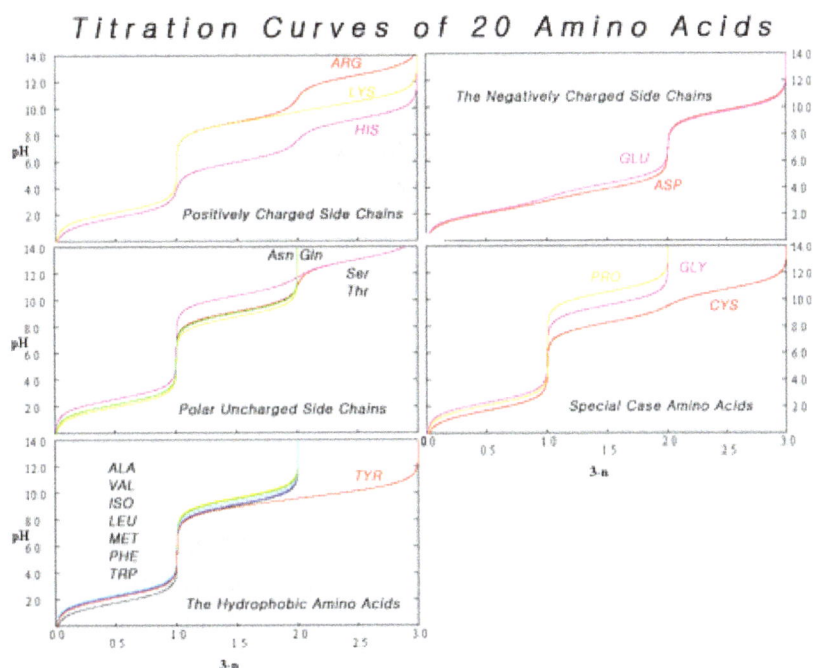

Composite of titration curves of twenty proteinogenic amino acids grouped by side chain category.

At pH values between the two pKa values, the zwitterion predominates, but coexists in dynamic equilibrium with small amounts of net negative and net positive ions. At the exact midpoint between the two pKa values, the trace amount of net negative and trace of net positive ions exactly balance, so that average net charge of all forms present is zero. This pH is known as the isoelectric point pI, so pI = ½(pKa₁ + pKa₂). The individual amino acids all have slightly different pKa values, so have different isoelectric points. For amino acids with charged side chains, the pKa of the side chain is involved. Thus for Asp, Glu with negative side chains, pI = ½(pKa₁ + pKaᵣ), where pKaᵣ is the side chain pKa. Cysteine also has potentially negative side chain with pKaᵣ = 8.14, so pI should be calculated as for Asp and Glu, even though the side chain is not significantly charged at neutral pH. For His, Lys, and Arg with positive side chains, pI = ½(pKaᵣ + pKa₂). Amino acids have zero mobility in electrophoresis at their isoelectric point, although this behaviour is more usually exploited for peptides and proteins than single amino acids. Zwitterions have minimum solubility at their isoelectric point and some amino acids (in particular, with non-polar side chains) can be isolated by precipitation from water by adjusting the pH to the required isoelectric point.

Occurrence and Functions in Biochemistry

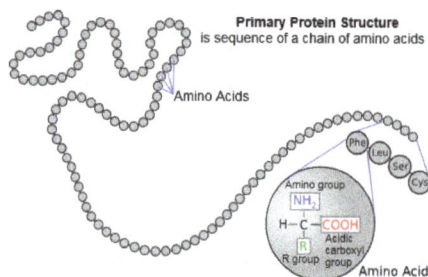

A polypeptide is an unbranched chain of amino acids.

Proteinogenic Amino Acids

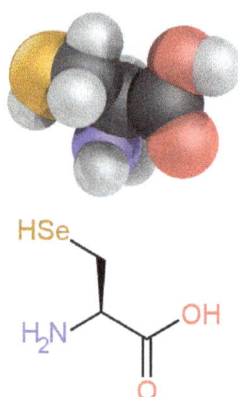

The amino acid selenocysteine

Amino acids are the structural units (monomers) that make up proteins. They join together to form short polymer chains called peptides or longer chains called either polypeptides or proteins. These polymers are linear and unbranched, with each amino acid within the chain attached to two neighboring amino acids. The process of making proteins is called *translation* and involves the step-by-step addition of amino acids to a growing protein chain by a ribozyme that is called a ribosome. The order in which the amino acids are added is read through the genetic code from an mRNA template, which is a RNA copy of one of the organism's genes.

Twenty-two amino acids are naturally incorporated into polypeptides and are called proteinogenic or natural amino acids. Of these, 20 are encoded by the universal genetic code. The remaining 2, selenocysteine and pyrrolysine, are incorporated into proteins by unique synthetic mechanisms. Selenocysteine is incorporated when the mRNA being translated includes a SECIS element, which causes the UGA codon to encode selenocysteine instead of a stop codon. Pyrrolysine is used by some methanogenic archaea in enzymes that they use to produce methane. It is coded for with the codon UAG, which is normally a stop codon in other organisms. This UAG codon is followed by a PYLIS downstream sequence.

Non-proteinogenic Amino Acids

Aside from the 22 proteinogenic amino acids, many *non-proteinogenic* amino acids are known. Those either are not found in proteins (for example carnitine, GABA, Levothyroxine) or are not produced directly and in isolation by standard cellular machinery (for example, hydroxyproline and selenomethionine).

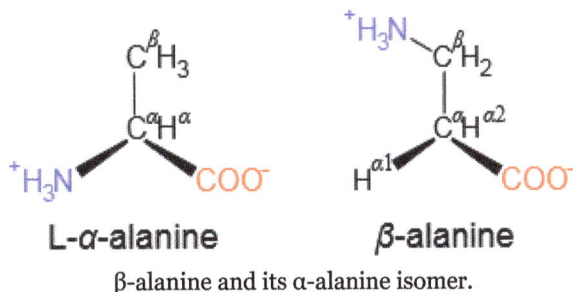

β-alanine and its α-alanine isomer.

Non-proteinogenic amino acids that are found in proteins are formed by post-translational modification, which is modification after translation during protein synthesis. These modifications are often essential for the function or regulation of a protein. Ffor example, the carboxylation of glutamate allows for better binding of calcium cations. connective tissues is composed of hydroxyproline, generated by hydroxylation of proline. Another example is the formation of hypusine in the translation initiation factor EIF5A, through modification of a lysine residue. Such modifications can also determine the localization of the protein, e.g., the addition of long hydrophobic groups can cause a protein to bind to a phospholipid membrane.

Some non-proteinogenic amino acids are not found in proteins. Examples include 2-aminoisobutyric acid and the neurotransmitter gamma-aminobutyric acid. Non-proteinogenic amino acids often occur as intermediates in the metabolic pathways for standard amino acids – for example, ornithine and citrulline occur in the urea cycle, part of amino acid catabolism. A rare exception to the dominance of α-amino acids in biology is the β-amino acid beta alanine (3-aminopropanoic acid), which is used in plants and microorganisms in the synthesis of pantothenic acid (vitamin B$_5$), a component of coenzyme A.

D-amino Acid Natural Abundance

D-isomers are uncommon in live organisms. For instance, gramicidin is a polypeptide made up from mixture of D- and L-amino acids. Other compounds containing D-amino acid are tyrocidine and valinomycin. These compounds disrupt bacterial cell walls, particularly in Gram-positive bacteria. Only 837 D-amino acids were found in Swiss-Prot database (187 million amino acids analysed).

Non-standard Amino Acids

The 20 amino acids that are encoded directly by the codons of the universal genetic code are called *standard* or *canonical* amino acids. A modified form of methionine (*N*-formylmethionine) is often incorporated in place of methionine as the initial amino acid of proteins in bacteria, mitochondria and chloroplasts. Other amino acids are called *non-standard* or *non-canonical*. Most of the non-standard amino acids are also non-proteinogenic (i.e. they cannot be incorporated into proteins during translation), but two of them are proteinogenic, as they can be incorporated translationally into proteins by exploiting information not encoded in the universal genetic code.

The two non-standard proteinogenic amino acids are selenocysteine (present in many non-eukaryotes as well as most eukaryotes, but not coded directly by DNA) and pyrrolysine (found only in some archaea and one bacterium). The incorporation of these non-standard amino acids is rare. For example, 25 human proteins include selenocysteine (Sec) in their primary structure, and the structurally characterized enzymes (selenoenzymes) employ Sec as the catalytic moiety in their active sites. Pyrrolysine and selenocysteine are encoded via variant codons. For example, seleno-cysteine is encoded by stop codon and SECIS element.

In Human Nutrition

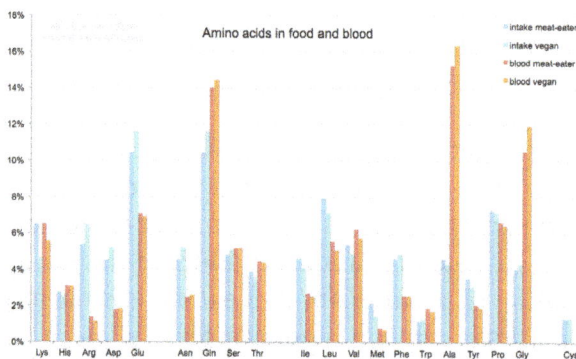

Share of amino acid in different human diets and the resulting mix of amino acids in human blood serum. Glutamate and glutamine are the most frequent in food at over 10%, while alanine, glutamine, and glycine are the most common in blood.

When taken up into the human body from the diet, the 20 standard amino acids either are used to synthesize proteins and other biomolecules or are oxidized to urea and carbon dioxide as a source of energy. The oxidation pathway starts with the removal of the amino group by a transaminase; the amino group is then fed into the urea cycle. The other product of transamidation is a keto acid that enters the citric acid cycle. Glucogenic amino acids can also be converted into glucose, through gluconeogenesis. Of the 20 standard amino acids, nine (His, Ile, Leu, Lys, Met, Phe, Thr, Trp and Val), are called essential amino acids because the human body cannot synthesize them from other compounds at the level needed for normal growth, so they must be obtained from food. In addition, cysteine, taurine, tyrosine, and arginine are considered semiessential amino-acids in children (though taurine is not technically an amino acid), because the metabolic pathways that synthesize these amino acids are not fully developed. The amounts required also depend on the age and health of the individual, so it is hard to make general statements about the dietary requirement for some amino acids. Dietary exposure to the non-standard amino acid BMAA has been linked to human neurodegenerative diseases, including ALS.

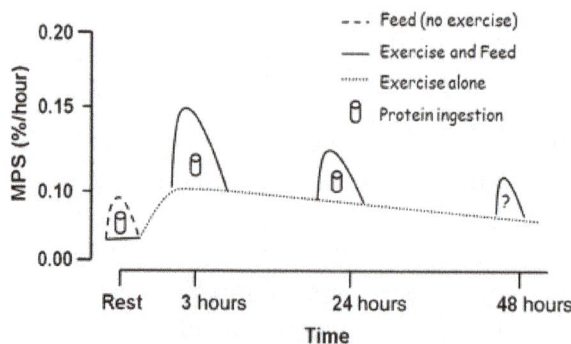

Resistance training stimulates muscle protein synthesis (MPS) for a period of up to 48 hours following exercise (shown by dotted line). Ingestion of a protein-rich meal at any point during this period will augment the exercise-induced increase in muscle protein synthesis (shown by solid lines).

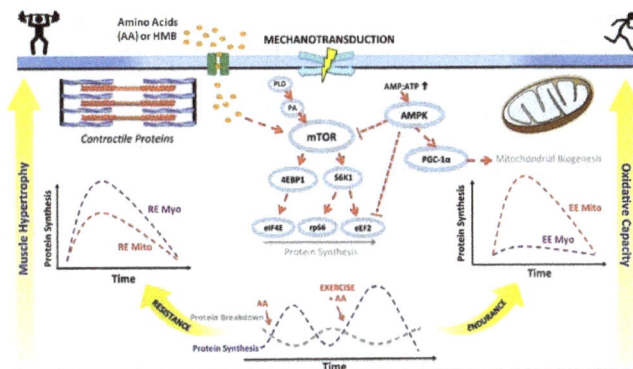

Diagram of the molecular signaling cascades that are involved in myofibrillar muscle protein synthesis and mitochondrial biogenesis in response to physical exercise and specific amino acids or their derivatives (primarily L-leucine and HMB). Many amino acids derived from food protein promote the activation of mTORC1 and increase protein synthesis by signaling through Rag GTPases.

Non-protein Functions

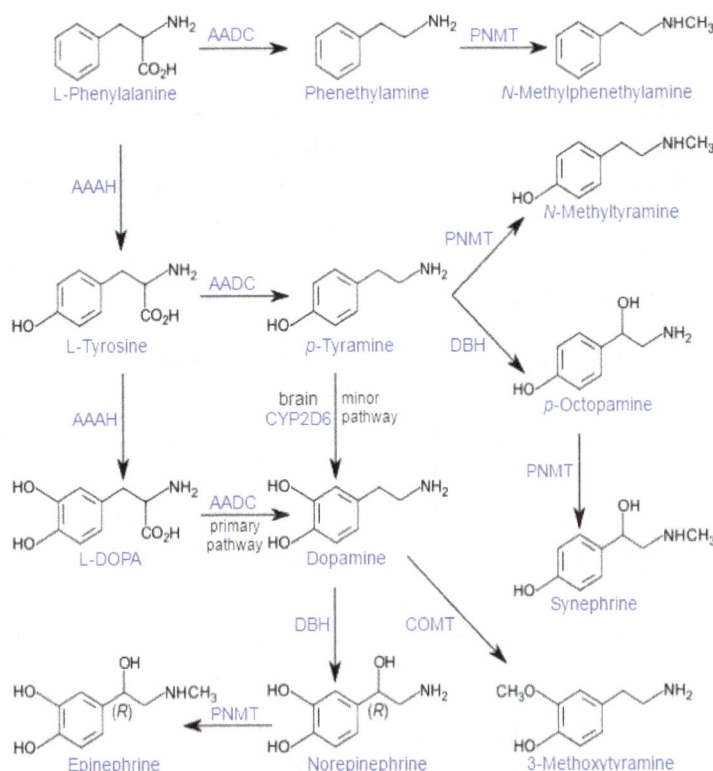

Catecholamines and trace amines are synthesized from phenylalanine and tyrosine in humans.

In humans, non-protein amino acids also have important roles as metabolic intermediates, such as in the biosynthesis of the neurotransmitter gamma-amino-butyric acid (GABA). Many amino acids are used to synthesize other molecules, for example:

- Tryptophan is a precursor of the neurotransmitter serotonin.

- Tyrosine (and its precursor phenylalanine) are precursors of the catecholamine neurotransmitters dopamine, epinephrine and norepinephrine and various trace amines.

- Phenylalanine is a precursor of phenethylamine and tyrosine in humans. In plants, it is a precursor of various phenylpropanoids, which are important in plant metabolism.

- Glycine is a precursor of porphyrins such as heme.

- Arginine is a precursor of nitric oxide.

- Ornithine and S-adenosylmethionine are precursors of polyamines.

- Aspartate, glycine, and glutamine are precursors of nucleotides. However, not all of the functions of other abundant non-standard amino acids are known.

Some non-standard amino acids are used as defenses against herbivores in plants. For example, canavanine is an analogue of arginine that is found in many legumes, and in particularly large amounts in *Canavalia gladiata* (sword bean). This amino acid protects the plants from predators such as insects and can cause illness in people if some types of legumes are eaten without processing. The non-protein amino acid mimosine is found in other species of legume, in particular *Leucaena leucocephala*. This compound is an analogue of tyrosine and can poison animals that graze on these plants.

Uses in Industry

Amino acids are used for a variety of applications in industry, but their main use is as additives to animal feed. This is necessary, since many of the bulk components of these feeds, such as soybeans, either have low levels or lack some of the essential amino acids: lysine, methionine, threonine, and tryptophan are most important in the production of these feeds. In this industry, amino acids are also used to chelate metal cations in order to improve the absorption of minerals from supplements, which may be required to improve the health or production of these animals.

The food industry is also a major consumer of amino acids, in particular, glutamic acid, which is used as a flavor enhancer, and aspartame (aspartyl-phenylalanine-1-methyl ester) as a low-calorie artificial sweetener. Similar technology to that used for animal nutrition is employed in the human nutrition industry to alleviate symptoms of mineral deficiencies, such as anemia, by improving mineral absorption and reducing negative side effects from inorganic mineral supplementation.

The chelating ability of amino acids has been used in fertilizers for agriculture to facilitate the delivery of minerals to plants in order to correct mineral deficiencies, such as iron chlorosis. These fertilizers are also used to prevent deficiencies from occurring and improving the overall health of the plants. The remaining production of amino acids is used in the synthesis of drugs and cosmetics.

Similarly, some amino acids derivatives are used in pharmaceutical industry. They include 5-HTP (5-hydroxytryptophan) used for experimental treatment of depression, L-DOPA (L-dihydroxyphenylalanine) for Parkinson's treatment, and eflornithine drug that inhibits ornithine decarboxylase and used in the treatment of sleeping sickness.

Expanded Genetic Code

Since 2001, 40 non-natural amino acids have been added into protein by creating a unique codon (recoding) and a corresponding transfer-RNA:aminoacyl – tRNA-synthetase pair to encode it with diverse physicochemical and biological properties in order to be used as a tool to exploring protein structure and function or to create novel or enhanced proteins.

Nullomers

Nullomers are codons that in theory code for an amino acid, however in nature there is a selective bias against using this codon in favor of another, for example bacteria prefer to use CGA instead of AGA to code for arginine. This creates some sequences that do not appear in the genome. This characteristic can be taken advantage of and used to create new selective cancer-fighting drugs and to prevent cross-contamination of DNA samples from crime-scene investigations.

Chemical Building Blocks

Amino acids are important as low-cost feedstocks. These compounds are used in chiral pool synthesis as enantiomerically pure building-blocks.

Amino acids have been investigated as precursors chiral catalysts, e.g., for asymmetric hydrogenation reactions, although no commercial applications exist.

Biodegradable Plastics

Amino acids are under development as components of a range of biodegradable polymers. These materials have applications as environmentally friendly packaging and in medicine in drug delivery and the construction of prosthetic implants. These polymers include polypeptides, polyamides, polyesters, polysulfides, and polyurethanes with amino acids either forming part of their main chains or bonded as side chains. These modifications alter the physical properties and reactivities of the polymers. An interesting example of such materials is polyaspartate, a water-soluble biodegradable polymer that may have applications in disposable diapers and agriculture. Due to its solubility and ability to chelate metal ions, polyaspartate is also being used as a biodegradeable anti-scaling agent and a corrosion inhibitor. In addition, the aromatic amino acid tyrosine is being developed as a possible replacement for toxic phenols such as bisphenol A in the manufacture of polycarbonates.

Reactions

As amino acids have both a primary amine group and a primary carboxyl group, these chemicals can undergo most of the reactions associated with these functional groups. These include nucleophilic addition, amide bond formation, and imine formation for the amine group, and esterification, amide bond formation, and decarboxylation for the carboxylic acid group. The combination of these functional groups allow amino acids to be effective polydentate ligands for metal-amino acid chelates. The multiple side chains of amino acids can also undergo chemical reactions. The types of these reactions are determined by the groups on these side chains and are, therefore, different between the various types of amino acid.

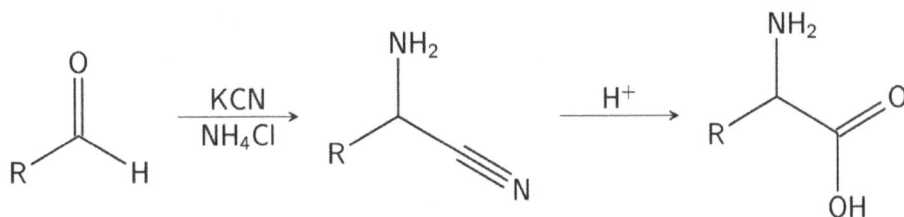

The Strecker amino acid synthesis

Chemical Synthesis

Several methods exist to synthesize amino acids. One of the oldest methods begins with the bromination at the α-carbon of a carboxylic acid. Nucleophilic substitution with ammonia then converts the alkyl bromide to the amino acid. In alternative fashion, the Strecker amino acid synthesis involves the treatment of an aldehyde with potassium cyanide and ammonia, this produces an α-amino nitrile as an intermediate. Hydrolysis of the nitrile in acid then yields a α-amino acid. Using ammonia or ammonium salts in this reaction gives unsubstituted amino acids, whereas substituting primary and secondary amines will yield substituted amino acids. Likewise, using ketones, instead of aldehydes, gives α,α-disubstituted amino acids. The classical synthesis gives racemic mixtures of α-amino acids as products, but several alternative procedures using asymmetric auxiliaries or asymmetric catalysts have been developed.

At the current time, the most-adopted method is an automated synthesis on a solid support (e.g., polystyrene beads), using protecting groups (e.g., Fmoc and t-Boc) and activating groups (e.g., DCC and DIC).

Peptide Bond Formation

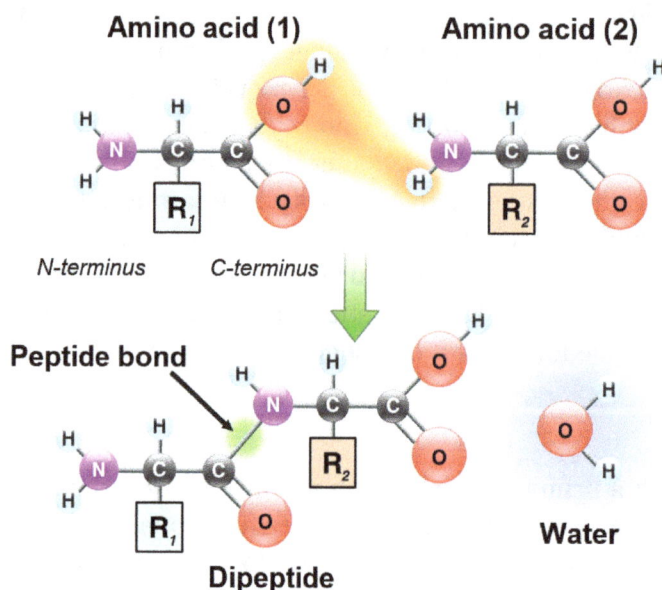

The condensation of two amino acids to form a *dipeptide* through a *peptide bond*

As both the amine and carboxylic acid groups of amino acids can react to form amide bonds, one amino acid molecule can react with another and become joined through an amide linkage. This polymerization of amino acids is what creates proteins. This condensation reaction yields the newly formed peptide bond and a molecule of water. In cells, this reaction does not occur directly; instead, the amino acid is first activated by attachment to a transfer RNA molecule through an ester bond. This aminoacyl-tRNA is produced in an ATP-dependent reaction carried out by an aminoacyl tRNA synthetase. This aminoacyl-tRNA is then a substrate for the ribosome, which catalyzes the attack of the amino group of the elongating protein chain on the ester bond. As a result of this mechanism, all proteins made by ribosomes are synthesized starting at their N-terminus and moving toward their C-terminus.

However, not all peptide bonds are formed in this way. In a few cases, peptides are synthesized by specific enzymes. For example, the tripeptide glutathione is an essential part of the defenses of cells against oxidative stress. This peptide is synthesized in two steps from free amino acids. In the first step, gamma-glutamylcysteine synthetase condenses cysteine and glutamic acid through a peptide bond formed between the side chain carboxyl of the glutamate (the gamma carbon of this side chain) and the amino group of the cysteine. This dipeptide is then condensed with glycine by glutathione synthetase to form glutathione.

In chemistry, peptides are synthesized by a variety of reactions. One of the most-used in solid-phase peptide synthesis uses the aromatic oxime derivatives of amino acids as activated units. These are added in sequence onto the growing peptide chain, which is attached to a solid resin support. The ability to easily synthesize vast numbers of different peptides by varying the types and order of amino acids (using combinatorial chemistry) has made peptide synthesis particularly important in creating libraries of peptides for use in drug discovery through high-throughput screening.

Biosynthesis

In plants, nitrogen is first assimilated into organic compounds in the form of glutamate, formed from alpha-ketoglutarate and ammonia in the mitochondrion. In order to form other amino acids, the plant uses transaminases to move the amino group to another alpha-keto carboxylic acid. For example, aspartate aminotransferase converts glutamate and oxaloacetate to alpha-ketoglutarate and aspartate. Other organisms use transaminases for amino acid synthesis, too.

Nonstandard amino acids are usually formed through modifications to standard amino acids. For example, homocysteine is formed through the transsulfuration pathway or by the demethylation of methionine via the intermediate metabolite S-adenosyl methionine, while hydroxyproline is made by a posttranslational modification of proline.

Microorganisms and plants can synthesize many uncommon amino acids. For example, some microbes make 2-aminoisobutyric acid and lanthionine, which is a sulfide-bridged derivative of alanine. Both of these amino acids are found in peptidic lantibiotics such as alamethicin. However, in plants, 1-aminocyclopropane-1-carboxylic acid is a small disubstituted cyclic amino acid that is a key intermediate in the production of the plant hormone ethylene.

Catabolism

Amino acids must first pass out of organelles and cells into blood circulation via amino acid transporters, since the amine and carboxylic acid groups are typically ionized. Degradation of an amino acid, occurring in the liver and kidneys, often involves deamination by moving its amino group to alpha-ketoglutarate, forming glutamate. This process involves transaminases, often the same as those used in amination during synthesis. In many vertebrates, the amino group is then removed through the urea cycle and is excreted in the form of urea. However, amino acid degradation can produce uric acid or ammonia instead. For example, serine dehydratase converts serine to pyruvate and ammonia. After removal of one or more amino groups, the remainder of the molecule can sometimes be used to synthesize new amino acids, or it can be used for energy by entering glycolysis or the citric acid cycle.

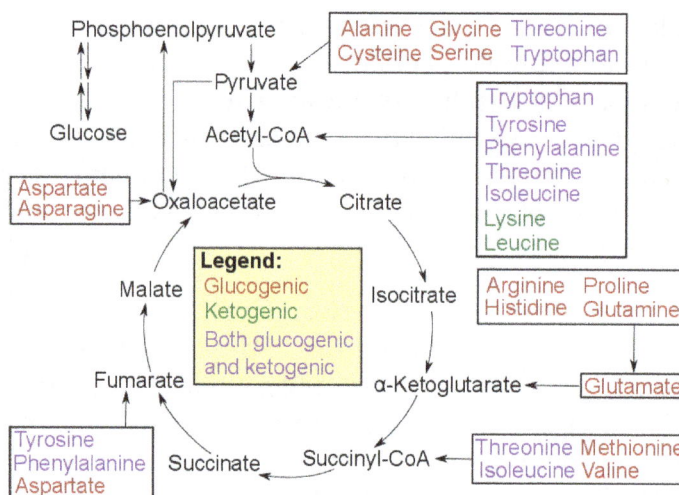

Catabolism of proteinogenic amino acids. Amino acids can be classified according to the properties of their main products as either of the following:

* Glucogenic, with the products having the ability to form glucose by gluconeogenesis
* Ketogenic, with the products not having the ability to form glucose. These products may still be used for ketogenesis or lipid synthesis.
* Amino acids catabolized into both glucogenic and ketogenic products.

Physicochemical Properties of Amino Acids

The 20 amino acids encoded directly by the genetic code can be divided into several groups based on their properties. Important factors are charge, hydrophilicity or hydrophobicity, size, and functional groups. These properties are important for protein structure and protein–protein interactions. The water-soluble proteins tend to have their hydrophobic residues (Leu, Ile, Val, Phe, and Trp) buried in the middle of the protein, whereas hydrophilic side chains are exposed to the aqueous solvent. (Note that in biochemistry, a residue refers to a specific monomer within the polymeric chain of a polysaccharide, protein or nucleic acid.) The integral membrane proteins tend to have outer rings of exposed hydrophobic amino acids that anchor them into the lipid bilayer. In the case part-way between these two extremes, some peripheral membrane proteins have a patch of hydrophobic amino acids on their surface that locks onto the membrane. In similar fashion, proteins that have to bind to positively charged molecules have surfaces rich with negatively charged amino acids like glutamate and aspartate, while proteins binding to negatively charged molecules have surfaces rich with positively charged chains like lysine and arginine. There are different hydrophobicity scales of amino acid residues.

Some amino acids have special properties such as cysteine, that can form covalent disulfide bonds to other cysteine residues, proline that forms a cycle to the polypeptide backbone, and glycine that is more flexible than other amino acids.

Many proteins undergo a range of posttranslational modifications, when additional chemical groups are attached to the amino acids in proteins. Some modifications can produce hydrophobic lipoproteins, or hydrophilic glycoproteins. These type of modification allow the reversible targeting of a protein to a membrane. For example, the addition and removal of the fatty acid palmitic acid to cysteine residues in some signaling proteins causes the proteins to attach and then detach from cell membranes.

The Basic Structure of Amino Acids

- It differs only in the structure of the side chain (R-group).

- L-isomer is normally found in proteins.

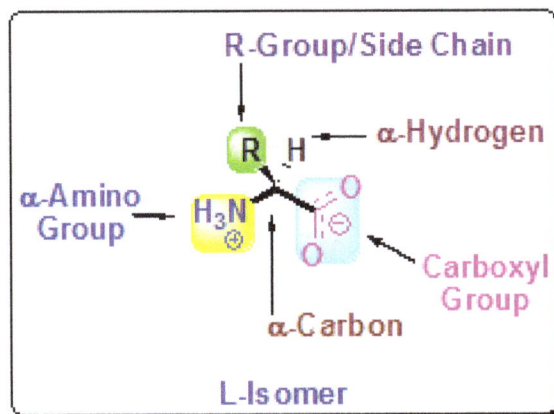

Nonionic and Zwitterionic Forms (Dipolar Structure) of Amino Acids

- Zwitterion = in German for hybrid ion

- The zwitterion predominates at neutral pH

- Isoelectric point (pI): pH at which the amino acid exists in a neutral, zwitterionic form (influenced by the nature of the side chain).

- At acidic pH, the carboxyl group is protonated and the amino acid is in the cationic form

- At neutral pH, the carboxyl group is deprotonated but the amino group is protonated. The net charge is zero; such ions are called Zwitterions

- At alkaline pH, the amino group is neutral $-NH_2$ and the amino acids are in the anionic form.

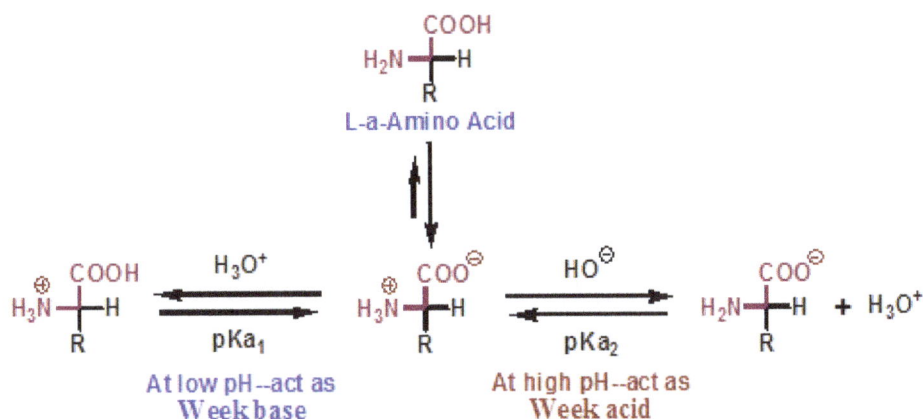

- Amino Acids Carry a Net Charge of Zero at a Specific pH.

> •Zwitterions predominate at pH values between the pK_a values of amino and carboxyl group
>
> •For amino acid without ionizable side chains, the Isoelectric Point (equivalence point, pI) is:
>
> $$pI = \frac{pK_1 + pK_2}{2}$$
>
> • At this point, the net charge is zero
>
>   – AA is least soluble in water
>
>   – AA does not migrate in electric field

- Amino acids have characteristic titration curves:

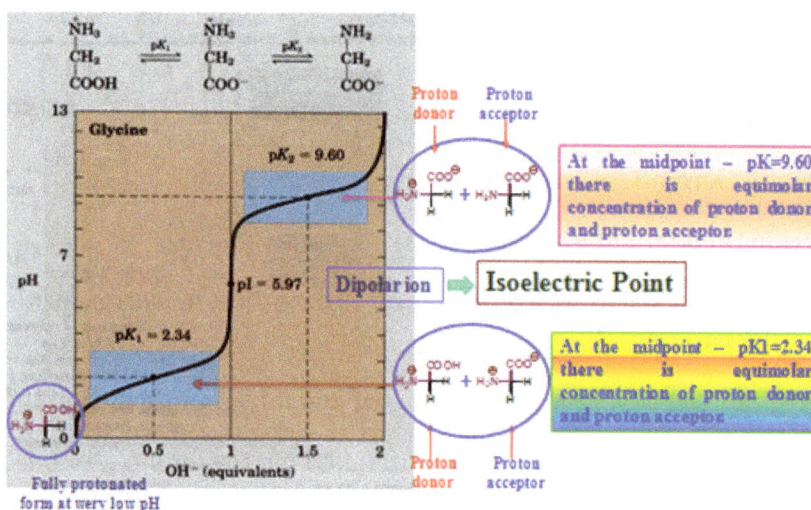

- Henderson/Hasselbach equation and pKa:

Protonated form   Unprotonated form (conjugate base)

$$HA \rightleftharpoons H^+ + A^-$$

$$K_a = \frac{[H^+][A^-]}{[HA]}$$

$$[H^+] = K_a \times \frac{[HA]}{[A^-]}$$

$$-\log[H^+] = -\log K_a - \log\frac{[HA]}{[A^-]}$$

$$pH = pK_a + \log\frac{[A^-]}{[HA]}$$

- ## Amino Acids Can Act as Buffers

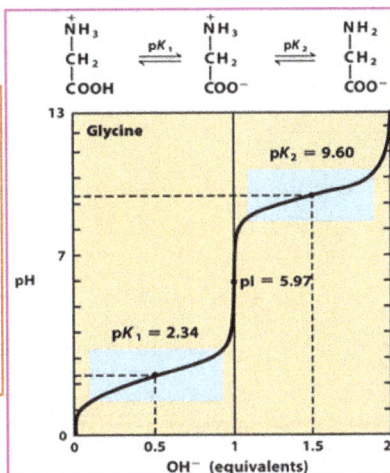

1. Amino acids with uncharged side-chains, such as glycine, have two pK_a values:
   - The pK_a of the α-carboxyl group is 2.34
   - The pK_a of the α-amino group is 9.6
2. Thus, it can act as a buffer in two pH regimes.

The Stereochemistry of Amino Acids

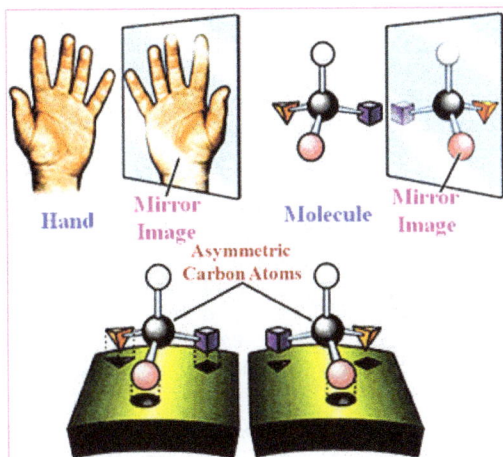

A. The Two Stereoisomers of Alanine

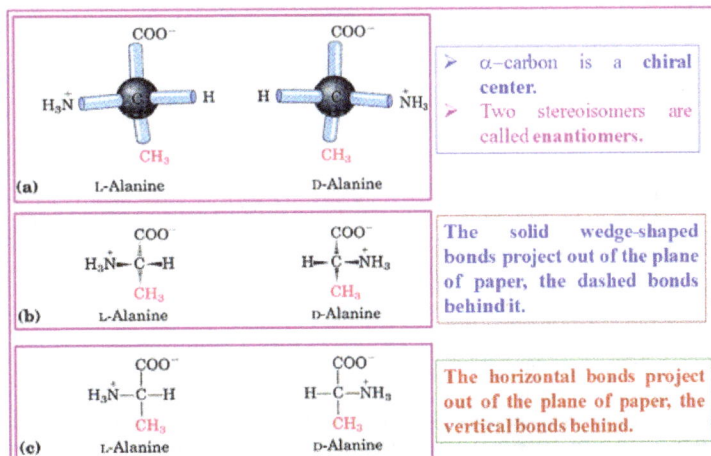

(a)   L-Alanine   D-Alanine

➤ α–carbon is a chiral center.
➤ Two stereoisomers are called enantiomers.

(b)   L-Alanine   D-Alanine

The solid wedge-shaped bonds project out of the plane of paper, the dashed bonds behind it.

(c)   L-Alanine   D-Alanine

The horizontal bonds project out of the plane of paper, the vertical bonds behind.

B. Stereoisomers of Threonine

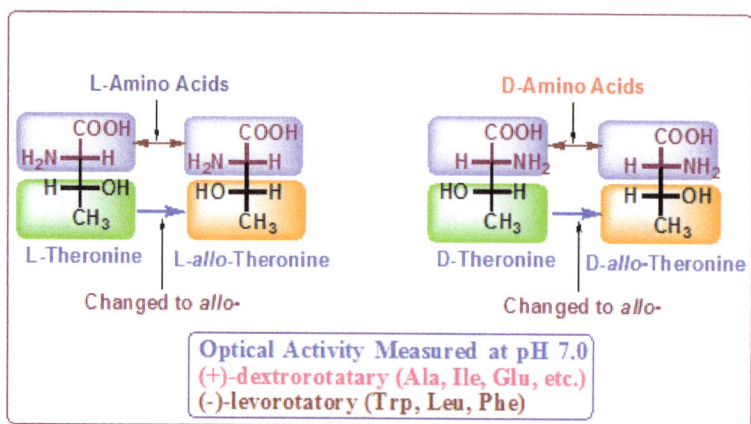

L-Amino Acids

D-Amino Acids

L-Theronine   L-allo-Theronine

D-Theronine   D-allo-Theronine

Changed to allo-   Changed to allo-

Optical Activity Measured at pH 7.0
(+)-dextrorotatary (Ala, Ile, Glu, etc.)
(-)-levorotatory (Trp, Leu, Phe)

Uncommon Amino Acids Found in Proteins

➢ Not incorporated by ribosomes

➢ Arise by post-translational modifications of proteins

➢ Reversible modifications, especially phosphorylation is important in regulation and signaling

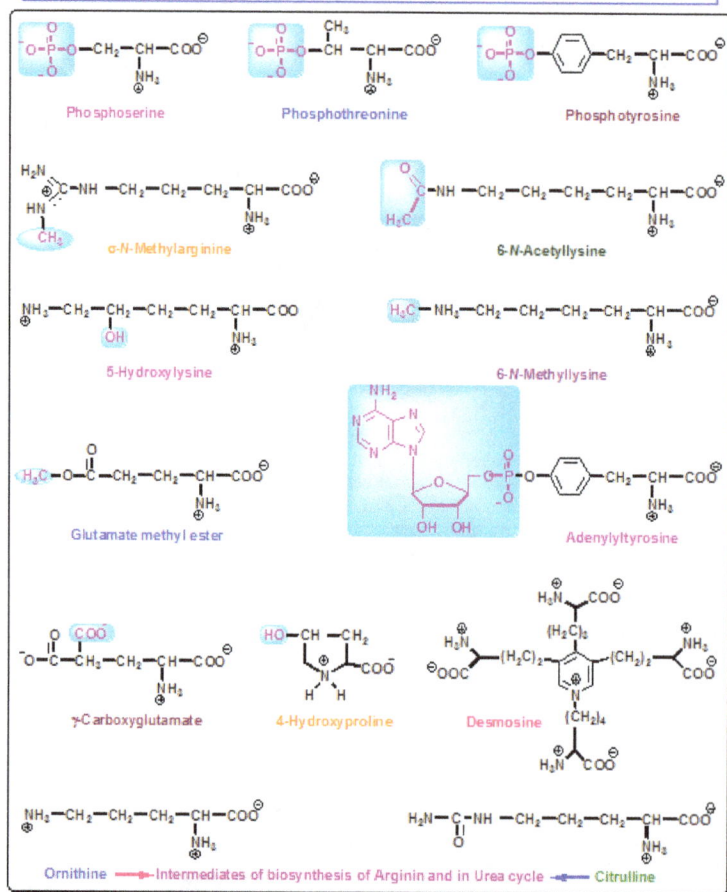

Phosphoserine   Phosphothreonine   Phosphotyrosine

α-N-Methylarginine   6-N-Acetyllysine

5-Hydroxylysine   6-N-Methyllysine

Glutamate methyl ester   Adenylyltyrosine

γ-Carboxyglutamate   4-Hydroxyproline   Desmosine

Ornithine ⟶ Intermediates of biosynthesis of Arginin and in Urea cycle ⟵ Citrulline

Structures of uncommon amino acids (post translational modification).

The Genetic Code: DNA-Amino Acid Dictionary

The Genetic Code: DNA-Amino Acid Dictionary

nonpolar	polar	basic	acidic	(stop codon)

1st base		2nd base			
		T	C	A	G
T	TTT (Phe/F) Phenylalanine	TCT (Ser/S) Serine	TAT (Tyr/Y) Tyrosine	TGT (Cys/C) Cysteine
	TTC (Phe/F) Phenylalanine	TCC (Ser/S) Serine	TAC (Tyr/Y) Tyrosine	TGC (Cys/C) Cysteine
	TTA (Leu/L) Leucine	TCA (Ser/S) Serine	TAA Ochre (Stop)	TGA Opal (Stop)
	TTG (Leu/L) Leucine	TCG (Ser/S) Serine	TAG Amber (Stop)	TGG (Trp/W) Tryptophan
C	CTT (Leu/L) Leucine	CCT (Pro/P) Proline	CAT (His/H) Histidine	CGT (Arg/R) Arginine
	CTC (Leu/L) Leucine	CCC (Pro/P) Proline	CAC (His/H) Histidine	CGC (Arg/R) Arginine
	CTA (Leu/L) Leucine	CCA (Pro/P) Proline	CAA (Gln/Q) Glutamine	CGA (Arg/R) Arginine
	CTG (Leu/L) Leucine	CCG (Pro/P) Proline	CAG (Gln/Q) Glutamine	CGG (Arg/R) Arginine
A	ATT (Ile/I) Isoleucine	ACT (Thr/T) Threonine	AAT (Asn/N) Asparagine	AGT (Ser/S) Serine
	ATC (Ile/I) Isoleucine	ACC (Thr/T) Threonine	AAC (Asn/N) Asparagine	AGC (Ser/S) Serine
	ATA (Ile/I) Isoleucine	ACA (Thr/T) Threonine	AAA (Lys/K) Lysine	AGA (Arg/R) Arginine
	ATG[A] (Met/M) Methionine	ACG (Thr/T) Threonine	AAG (Lys/K) Lysine	AGG (Arg/R) Arginine
G	GTT (Val/V) Valine	GCT (Ala/A) Alanine	GAT (Asp/D) Aspartic acid	GGT (Gly/G) Glycine
	GTC (Val/V) Valine	GCC (Ala/A) Alanine	GAC (Asp/D) Aspartic acid	GGC (Gly/G) Glycine
	GTA (Val/V) Valine	GCA (Ala/A) Alanine	GAA (Glu/E) Glutamic acid	GGA (Gly/G) Glycine
	GTG (Val/V) Valine	GCG (Ala/A) Alanine	GAG (Glu/E) Glutamic acid	GGG (Gly/G) Glycine

[A] The codon ATG both codes for methionine and serves as an initiation site: the first ATG in an DNA's coding region is where translation into protein begins. 21ˢᵗ AA, Selenocysteine, codes "UGA" opal (or umber) stop codon and 22ⁿᵈ AA, Pyrrolysine, Codes "GUA" amber stop codon

Synthesis of α-Amino Acids

New unnatural amino acids with altered properties

- new therapeutics (lead compounds)

- mechanistic probes

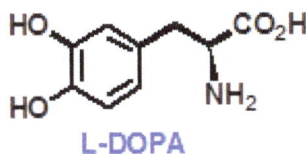

L-DOPA

Alpha and Beta Carbon

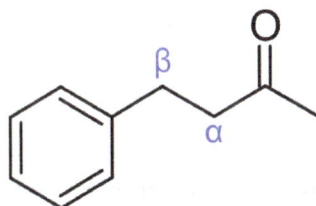

Alpha and beta carbons in a skeletal formula. The carbonyl has two β-hydrogens and five α-hydrogens

The alpha carbon (Cα) in organic molecules refers to the first carbon atom that attaches to a functional group, such as a carbonyl. The second carbon atom is called the beta carbon (Cβ), and the system continues naming in alphabetical order with Greek letters.

The nomenclature can also be applied to the hydrogen atoms attached to the carbon atoms. A hydrogen atom attached to an alpha carbon atom is called an alpha-hydrogen atom, a hydrogen atom on the beta-carbon atom is a beta hydrogen atom, and so on.

This naming standard may not be in compliance with IUPAC nomenclature, which encourages that carbons be identified by number, not by Greek letter, but it nonetheless remains very popular, in particular because it is useful in identifying the relative location of carbon atoms to other functional groups.

Organic molecules with more than one functional group can be a source of confusion. Generally the functional group responsible for the name or type of the molecule is the 'reference' group for purposes of carbon-atom naming. For example, the molecules nitrostyrene and phenethylamine are very similar; the former can even be reduced into the latter. However, nitrostyrene's α-carbon atom is adjacent to the styrene group; in phenethylamine this same carbon atom is the β-carbon atom, as phenethylamine (being an amine rather than a styrene) counts its atoms from the opposite "end" of the molecule.

Examples

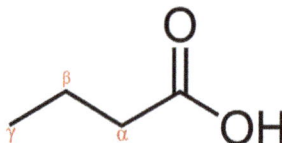

Skeletal formula of butyric acid with the alpha, beta, and gamma carbons marked

Proteins and Amino Acids

Alpha-carbon (α-carbon) is also a term that applies to proteins and amino acids. It is the backbone carbon before the carbonyl carbon. Therefore, reading along the backbone of a typical protein would give a sequence of $-[N-C\alpha-\text{carbonyl C}]_n-$ etc. (when reading in the N to C direction). The α-carbon is where the different substituents attach to each different amino acid. That is, the groups hanging off the chain at the α-carbon are what give amino acids their diversity. These groups give the α-carbon its stereogenic properties for every amino acid except for glycine. Therefore, the α-carbon is a stereocenter for every amino acid except glycine. Glycine also does not have a β-carbon, while every other amino acid does.

The α-carbon of an amino acid is significant in protein folding. When describing a protein, which is a chain of amino acids, one often approximates the location of each amino acid as the location of its α-carbon. In general, α-carbons of adjacent amino acids in a protein are about 3.8 ångströms (380 picometers) apart.

Enols and Enolates

The α-carbon is important for enol- and enolate-based carbonyl chemistry as well. Chemical transformations affected by the conversion to either an enolate or an enol, in general, lead to the α-carbon acting as a nucleophile, becoming, for example, alkylated in the presence of primary haloalkane. An exception is in reaction with silyl-chlorides, -bromides, and -iodides, where the oxygen acts as the nucleophile to produce silyl enol ether.

Non-proteinogenic Amino Acids

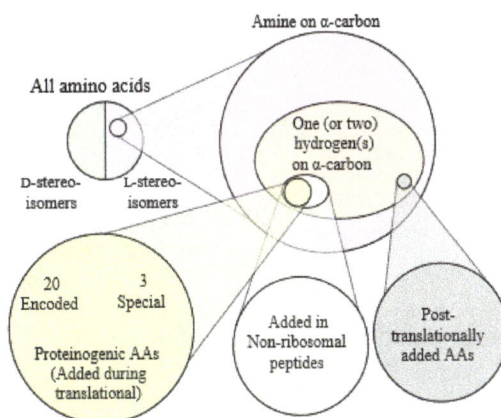

Proteinogenic amino acids are a small fraction of all amino acids

In biochemistry, non-coded or non-proteinogenic amino acids are those not naturally encoded or found in the genetic code of any organism. Despite the use of only 22 amino acids (21 in eukaryotes) by the translational machinery to assemble proteins (the proteinogenic amino acids), over 140 amino acids are known to occur naturally in proteins and thousands more may occur in nature or be synthesized in the laboratory. Many non-proteinogenic amino acids are noteworthy because they are:

- intermediates in biosynthesis,

- post-translationally formed in proteins,

- possess a physiological role (e.g. components of bacterial cell walls, neurotransmitters and toxins),

- natural or man-made pharmacological compounds,

- present in meteorites and in prebiotic experiments (e.g. Miller–Urey experiment).

Definition by Negation

Technically, any organic compound with an amine (-NH_2) and a carboxylic acid (-COOH) functional group is an amino acid. The proteinogenic amino acids are small subset of this group that possess central carbon atom (α- or 2-) bearing an amino group, a carboxyl group, a side chain and an α-hydrogen levo conformation, with the exception of glycine, which is achiral, and proline, whose amine group is a secondary amine and is consequently frequently referred to as an imino acid for traditional reasons, albeit not an imino.

The genetic code encodes 20 standard amino acids for incorporation into proteins during translation. However, there are two extra proteinogenic amino acids: selenocysteine and pyrrolysine. These non-standard amino acids do not have a dedicated codon, but are added in place of a stop codon when a specific sequence is present, UGA codon and SECIS element for selenocysteine, UAG PYLIS downstream sequence for pyrrolysine. All other amino acids are termed "non-proteinogenic".

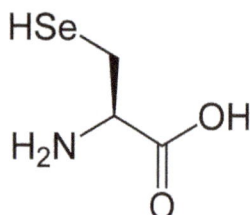

Selenocysteine. This amino acid contains a selenol group on its β-carbon

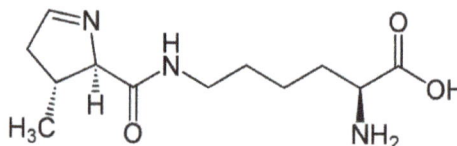

Pyrrolysine. This amino acid is formed by joining to the ε-amino group of lysine a carboxylated pyrroline ring

There are various groups of amino acids:

- 20 standard amino acids

- 22 proteinogenic amino acids

- over 80 amino acids created abiotically in high concentrations

- about 900 are produced by natural pathways

- over 118 engineered amino acids have been placed into protein

These groups overlap, but are not identical. All 22 proteinogenic amino acids are biosynthesised by organisms and some, but not all, of them also are abiotic (found in prebiotic experiments and meteorites). Some natural amino acids, such as norleucine, are misincorporated translationally into proteins due to infidelity of the protein-synthesis process. Many amino acids, such as ornithine, are metabolic intermediates produced biosynthetically, but not incorporated translationally into proteins. Post-translational modification of amino-acid residues in proteins leads to the formation of many proteinaceous, but non-proteinogenic, amino acids. Other amino acids are solely found in abiotic mixes (e.g. α-methylnorvaline). Over 30 unnatural amino acids have been inserted translationally into protein in engineered systems, yet are not biosynthetic.

Nomenclature

In addition to the IUPAC numbering system to differentiate the various carbons in an organic molecule, by sequentially assigning a number to each carbon, including those forming a carboxylic group, the carbons along the side-chain of amino acids can also be labelled with Greek letters, where the α-carbon is the central chiral carbon possessing a carboxyl group, a side chain and, in α-amino acids, an amino group – the carbon in carboxylic groups is not counted. (Consequently, the IUPAC names of many non-proteinogenic α-amino acids start with *2-amino-* and end in *-ic acid*.)

Natural, but Non L-α-amino Acids

Most natural amino acids are α-amino acids in the L conformation, but some exceptions exist.

Non-alpha

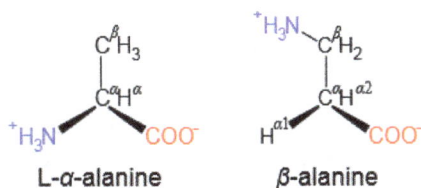

L-α-alanine   β-alanine

Some non-α amino acids exist in organisms. In these structures, the amine group displaced further from the carboxylic acid end of the amino acid molecule. Thus a β amino acid has the amine group bonded to the second carbon away, and a γ amino acid has it on the third. Examples include β-alanine, GABA, and δ-aminolevulinic acid.

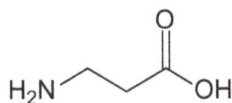

- β-alanine: an amino acid produced by aspartate 1-decarboxylase and a precursor to coenzyme

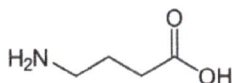

- γ-Aminobutyric acid (GABA): a neurotransmitter in animals.

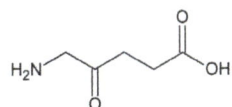

- δ-Aminolevulinic acid: an intermediate in tetrapyrrole biosynthesis (haem, chlorophyll, cobalamin *etc.*).

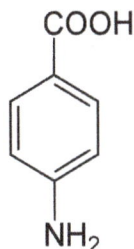

- 4-Aminobenzoic acid (PABA): an intermediate in folate biosynthesis

The reason why α-amino acids are used in proteins has been linked to their frequency in meteorites and prebiotic experiments. An initial speculation on the deleterious properties of β-amino acids in terms of secondary structure, turned out to be incorrect.

D-amino Acids

Some amino acids contain the opposite absolute chirality, chemicals that are not available from normal ribosomal translation/transcription machinery. Most bacterial cells walls are formed by peptidoglycan, a polymer composed of amino sugars crosslinked with short oligopeptides bridged between each other. The oligopeptide is non-ribosomally synthesised and contains several peculiarities, including D-amino acids, generally D-alanine and D-glutamate. A further peculiarity is that the former is racemised by a PLP-binding enzymes (encoded by *alr* or the homologue *dadX*), whereas the latter is racemised by a cofactor independent enzyme (*murI*). Some variants are present, in *Thermotoga* spp. D-lysine is present and in certain vancomycin-resistant bacteria D-serine is present (*vanT* gene).

In animals, some D-amino acids are neurotransmitters.

Without a Hydrogen on the α-carbon

All proteinogenic amino acids have at least one hydrogen on the α-carbon. Glycine has two hydrogens, and all others have one hydrogen and one side-chain. Replacement of the remaining hydrogen with a larger substituent, such as a methyl group, distorts the protein backbone.

In some fungi α-amino isobutyric acid is produced as a precursor to peptides, some of which exhibit antibiotic properties. This compound is similar to alanine, but possesses an additional methyl group on the α-carbon instead of a hydrogen. It is therefore achiral. Another compound similar to alanine without an α-hydrogen is dehydroalanine, which possess a methylene sidechain. It is one of several naturally occurring dehydroamino acids.

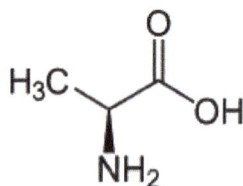

- alanine

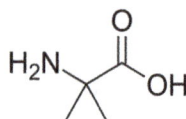

- aminoisobutyric acid

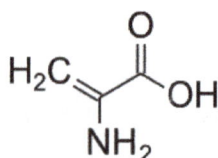

- dehydroalanine

Twin Amino Acid Stereocentres

A subset of L-α-amino acids are ambiguous as to which of two ends is the α-carbon. In proteins a cysteine residue can form a disulfide bond with another cysteine residue, thus crosslinking the protein. Two crosslinked cysteines form a cystine molecule. Cysteine and methionine are generally produced by direct sulfurylation, but in some species they can be produced by transsulfuration, where the activated homoserine or serine is fused to a cysteine or homocysteine forming cystathionine. A similar compound is lanthionine, which can be seen as two alanine molecules joined via a thioether bond and is found in various organisms. Similarly, djenkolic acid, a plant toxin from jengkol beans, is composed of two cysteines connected by a methylene group. Diaminopimelic acid is both used as a bridge in petidoglycan and is used a precursor to lysine (via its decarboxylation).

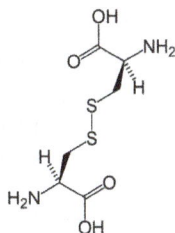

- cystine

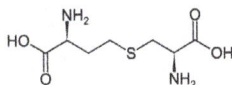

- cystathionine

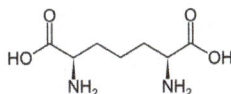

- lanthionine

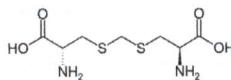

- Djenkolic acid

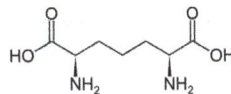

- Diaminopimelic acid

Prebiotic Amino Acids and Alternative Biochemistries

In meteorites and in prebiotic experiments (e.g. Miller–Urey experiment) many more amino acids than the twenty standard amino acids are found, several of which at higher concentrations that the standard ones: it has been conjectured that if amino acid based life were to arise in parallel elsewhere in the universe, no more than 75% of the amino acids would be in common. The most notable anomaly is the lack of aminobutyric acid.

Proportion of amino acids relative to glycine (%)		
Molecule	Electric discharge	Murchinson meteorite
Glycine	100	100
Alanine	180	36
α-Amino-n-butyric acid	61	19
Norvaline	14	14
Valine	4.4	
Norleucine	1.4	
Leucine	2.6	
Isoleucine	1.1	
Alloisoleucine	1.2	
t-leucine	< 0.005	
α-Amino-n-heptanoic acid	0.3	
Proline	0.3	22
Pipecolic acid	0.01	11
α,β-diaminopropionic acid	1.5	
α,γ-diaminobutyric acid	7.6	
Ornithine	< 0.01	
lysine	< 0.01	
Aspartic acid	7.7	13
Glutamic acid	1.7	20
Serine	1.1	
Threonine	0.2	
Allothreonine	0.2	
Methionine	0.1	
Homocysteine	0.5	
Homoserine	0.5	
β-Alanine	4.3	10
β-Amino-n-butyric acid	0.1	5
β-Aminoisobutyric acid	0.5	7
γ-Aminobutyric acid	0.5	7
α-Aminoisobutyric acid	7	33
isovaline	1	11
Sarcosine	12.5	7
N-ethyl glycine	6.8	6
N-propyl glycine	0.5	
N-isopropyl glycine	0.5	

Proportion of amino acids relative to glycine (%)		
Molecule	Electric discharge	Murchinson meteorite
N-methyl alanine	3.4	3
N-ethyl alanine	< 0.05	
N-methyl β-alanine	1.0	
N-ethyl β-alanine	< 0.05	
isoserine	1.2	
α-hydroxy-γ-aminobutyric acid	17	

Straight Side Chain

The genetic code has been described as a frozen accident and the reasons why there is only one standard amino acid with a straight chain (alanine) could simply be redundancy with valine, leucine and isoleucine. However, straight chained amino acids are reported to form much more stable alpha helices.

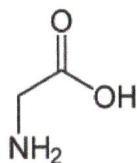

- Glycine (Hydrogen side-chain)

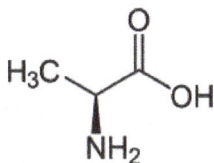

- Alanine (Methyl side-chain)

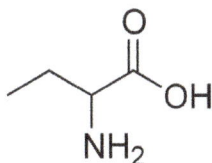

- α-aminobutyric acid (Ethyl side-chain)

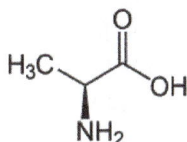

- Norvaline (*n*-Propyl side-chain)

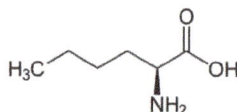

- Norleucine (*n*-Butyl side-chain)

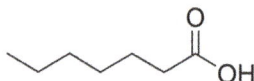

- Homonorleucine (*n*-Pentyl side-chain)

Chalcogen

Serine, homoserine, O-methyl-homoserine and O-ethyl-homoserine possess an hydroxymethyl, hydroxyethyl, O-methyl-hydroxymethyl and O-methyl-hydroxyethyl side chain. Whereas cysteine, homocysteine, methionine and ethionine possess the thiol equivalents. The selenol equivalents are selenocysteine, selenohomocysteine, selenomethionine and selenoethionine. Amino acids with the next chalcogen down are also found in nature: several species such as Aspergillus fumigatus, Aspergillus terreus, and Penicillium chrysogenum in the absence of sulfur are able to produce and incorporate into protein tellurocysteine and telluromethionine.

Hydroxyglycine, an amino acid with a hydroxyl side-chain, is highly unstable.

Expanded Genetic Code

Roles

In cells, especially autotrophs, several non-proteinogenic amino acids are found as metabolic intermediates. However, despite the catalytic flexibility of PLP-binding enzymes, many amino acids are synthesised as keto-acids (*e.g.* 4-methyl-2-oxopentanoate to leucine) and aminated in the last step, thus keeping the number of non-proteinogenic amino acid intermediates fairly low.

Ornithine and citrulline occur in the urea cycle, part of amino acid catabolism.

In addition to primary metabolism, several non-proteinogenic amino acids are precursors or the final production in secondary metabolism to make small compounds or non-ribosomal peptides (such as some toxins).

Post-translationally Incorporated into Protein

Despite not being encoded by the genetic code as proteinogenic amino acids, some non-standard amino acids are nevertheless found in proteins. These are formed by post-translational modification of the side chains of standard amino acids present in the target protein. These modifications are often essential for the function or regulation of a protein; for example, in Gamma-carboxyglutamate the carboxylation of glutamate allows for better binding of calcium cations, and in hydroxyproline the hydroxylation of proline is critical for maintaining connective tissues. Another example is the formation of hypusine in the translation initiation factor EIF5A, through modification of a lysine residue. Such modifications can also determine the localization of the protein, e.g., the addition of long hydrophobic groups can cause a protein to bind to a phospholipid membrane.

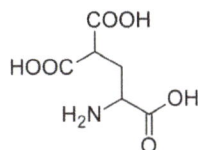

- Carboxyglutamic acid. Whereas glutamic acid possess one γ-carboxyl group, Carboxyglutamic acid possess two.

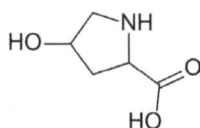

- Hydroxyproline. This imino acid differs from proline due to a hydroxyl group on carbon 4.

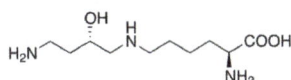

- Hypusine. This amino acid is obtained by adding to the ε-amino group of a lysine a 4-aminobutyl moiety (obtained from spermidine)

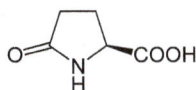

- Pyroglutamic acid

There is some preliminary evidence that aminomalonic acid may be present, possibly by misincorporation, in protein.

Toxic Analogues

Several non-proteinogenic amino acids are toxic due to their ability to mimic certain properties of proteinogenic amino acids, such as thialysine. Some non-proteinogenic amino acids are neurotoxic by mimicking amino acids used as neurotransmitters (i.e. not for protein biosynthesis), e.g. Quisqualic acid, canavanine or azetidine-2-carboxylic acid. Cephalosporin C has an α-aminoadipic acid (homoglutamate) backbone that is amidated with a cephalosporin moiety. Penicillamine is therapeutic amino acid, whose mode of action is unknown.

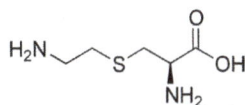

- Thialysine

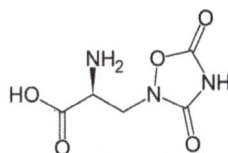

- Quisqualic acid

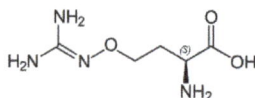

- Canavanine

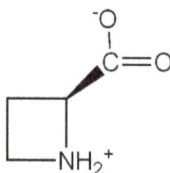

- azetidine-2-carboxylic acid

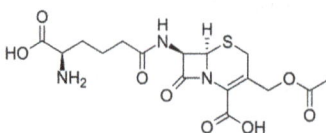

- Cephalosporin C

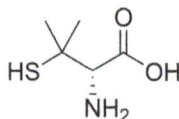

- Penicillamine

Naturally-occurring cyanotoxins can also include non-proteinogenic amino acids. Microcystin and nodularin, for example, are both derived from ADDA, a β-amino acid.

Not Amino Acids

Taurine is an amino sulfonic acid and not an amino acid, however it is occasionally considered as such as the amounts required to suppress the auxotroph in certain organisms (e.g. cats) are closer to those of "essential amino acids" (amino acid auxotrophy) than of vitamins (cofactor auxotrophy).

The osmolytes, sarcosine and glycine betaine are derived from amino acids, but have a secondary and quaternary amine respectively.

References

- Simoni RD, Hill RL, Vaughan M (September 2002). "The discovery of the amino acid threonine: the work of William C. Rose [classical article]". The Journal of Biological Chemistry. 277 (37): E25. PMID 12218068

- Pollegioni L, Servi S, eds. (2012). Unnatural Amino Acids: Methods and Protocols. Methods in Molecular Biology – Volume 794. Humana Press. p. v. ISBN 978-1-61779-331-8. OCLC 756512314

- "Chapter 1: Proteins are the Body's Worker Molecules". The Structures of Life. National Institute of General Medical Sciences. 27 October 2011. Retrieved 20 May 2008

- Wagner I, Musso H (November 1983). "New Naturally Occurring Amino Acids". Angewandte Chemie International Edition in English. 22 (11): 816–28. doi:10.1002/anie.198308161

- McCoy RH, Meyer CE, Rose WC (1935). "Feeding Experiments with Mixtures of Highly Purified Amino Acids. VIII. Isolation and Identification of a New Essential Amino Acid". Journal of Biological Chemistry. 112: 283–302

- Michal G, Schomburg D, eds. (2012). Biochemical Pathways: An Atlas of Biochemistry and Molecular Biology (2nd ed.). Oxford: Wiley-Blackwell. p. 5. ISBN 978-0-470-14684-2

- Hatem, Salama Mohamed Ali (2006). "Gas chromatographic determination of Amino Acid Enantiomers in tobacco and bottled wines". University of Giessen. Retrieved 17 November 2008

- Rodnina MV, Beringer M, Wintermeyer W (January 2007). "How ribosomes make peptide bonds". Trends in Biochemical Sciences. 32 (1): 20–6. PMID 17157507. doi:10.1016/j.tibs.2006.11.007

- Trown, P. W.; Smith, B.; Abraham, E. P. (1963). "Biosynthesis of cephalosporin C from amino acids". The Biochemical Journal. 86 (2): 284–291. PMC 1201751 . PMID 13994319. doi:10.1042/bj0860284

- Simon M (2005). Emergent computation: emphasizing bioinformatics. New York: AIP Press/Springer Science+Business Media. pp. 105–106. ISBN 0-387-22046-1

- "Nomenclature and Symbolism for Amino Acids and Peptides". IUPAC-IUB Joint Commission on Biochemical Nomenclature. 1983. Archived from the original on 9 October 2008. Retrieved 17 November 2008

- Blenis J, Resh MD (December 1993). "Subcellular localization specified by protein acylation and phosphorylation". Current Opinion in Cell Biology. 5 (6): 984–9. PMID 8129952. doi:10.1016/0955-0674(93)90081-Z

- Vermeer, C. (1990). "Gamma-carboxyglutamate-containing proteins and the vitamin K-dependent carboxylase". The Biochemical Journal. 266 (3): 625–636. PMC 1131186 . PMID 2183788. doi:10.1042/bj2660625

- Stryer L, Berg JM, Tymoczko JL (2002). Biochemistry (5th ed.). New York: W.H. Freeman. pp. 693–8. ISBN 978-0-7167-4684-3

- Baumann E (1884). "Über cystin und cystein". Z Physiol Chem. 8 (4): 299–305. Archived from the original on 14 March 2011. Retrieved 28 March 2011

- Kostrzewa RM, Nowak P, Kostrzewa JP, Kostrzewa RA, Brus R (March 2005). "Peculiarities of L: -DOPA treatment of Parkinson's disease". Amino Acids. 28 (2): 157–64. PMID 15750845. doi:10.1007/s00726-005-0162-4

- Padmanabhan, S.; Baldwin, R. L. (1991). "Straight-chain non-polar amino acids are good helix-formers in water". Journal of Molecular Biology. 219 (2): 135–137. PMID 2038048. doi:10.1016/0022-2836(91)90553-I

- Hausman, Robert E., Cooper, Geoffrey M. (2004). The cell: a molecular approach. Washington, D.C: ASM Press. p. 51. ISBN 0-87893-214-3

- Elzanowski A, Ostell J (7 April 2008). "The Genetic Codes". National Center for Biotechnology Information (NCBI). Retrieved 10 March 2010

Peptide: Bonds and Synthesis

The small chains formed by amino acids are linked by bonds formed by peptides. Peptides are smaller in size than proteins and usually consist of 2 to 50 amino acids. The topics elaborated in this chapter will help in gaining a better perspective on the subject of peptide bond and synthesis.

Peptide

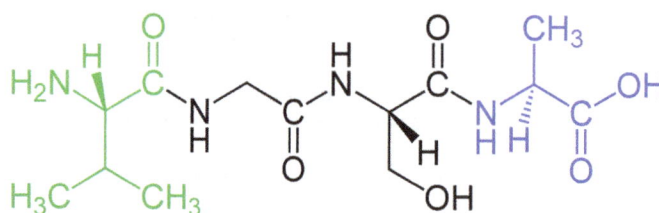

A tetrapeptide (example Val-Gly-Ser-Ala) with green marked amino end (L-Valine) and blue marked carboxyl end (L-Alanine).

Peptides ("digested"; derived from "to digest") are biologically occurring short chains of amino acid monomers linked by peptide (amide) bonds.

The covalent chemical bonds are formed when the carboxyl group of one amino acid reacts with the amine group of another. The shortest peptides are dipeptides, consisting of 2 amino acids joined by a single peptide bond, followed by tripeptides, tetrapeptides, etc. A polypeptide is a long, continuous, and unbranched peptide chain. Hence, peptides fall under the broad chemical classes of biological oligomers and polymers, alongside nucleic acids, oligosaccharides and polysaccharides, etc.

Peptides are distinguished from proteins on the basis of size, and as an arbitrary benchmark can be understood to contain approximately 50 or fewer amino acids. Proteins consist of one or more polypeptides arranged in a biologically functional way, often bound to ligands such as coenzymes and cofactors, or to another protein or other macromolecule (DNA, RNA, etc.), or to complex macromolecular assemblies. Finally, while aspects of the lab techniques applied to peptides versus polypeptides and proteins differ (e.g., the specifics of electrophoresis, chromatography, etc.), the size boundaries that distinguish peptides from polypeptides and proteins are not absolute: long peptides such as amyloid beta have been referred to as proteins, and smaller proteins like insulin have been considered peptides.

Amino acids that have been incorporated into peptides are termed "residues" due to the release of either a hydrogen ion from the amine end or a hydroxyl ion from the carboxyl end, or both, as a water molecule is released during formation of each amide bond. All peptides except cyclic peptides have an N-terminal and C-terminal residue at the end of the peptide (as shown for the tetrapeptide in the image).

Peptide Classes

Peptides are divided into several classes, depending on how they are produced:

Milk peptides

> Two naturally occurring milk peptides are formed from the milk protein casein when digestive enzymes break this down; they can also arise from the proteinases formed by lactobacilli during the fermentation of milk.

Ribosomal peptides

> Ribosomal peptides are synthesized by translation of mRNA. They are often subjected to proteolysis to generate the mature form. These function, typically in higher organisms, as hormones and signaling molecules. Some organisms produce peptides as antibiotics, such as microcins. Since they are translated, the amino acid residues involved are restricted to those utilized by the ribosome.

However, these peptides frequently have posttranslational modifications such as phosphorylation, hydroxylation, sulfonation, palmitoylation, glycosylation and disulfide formation. In general, they are linear, although lariat structures have been observed. More exotic manipulations do occur, such as racemization of L-amino acids to D-amino acids in platypus venom.

Nonribosomal peptides

> Nonribosomal peptides are assembled by enzymes that are specific to each peptide, rather than by the ribosome. The most common non-ribosomal peptide is glutathione, which is a component of the antioxidant defenses of most aerobic organisms. Other nonribosomal peptides are most common in unicellular organisms, plants, and fungi and are synthesized by modular enzyme complexes called *nonribosomal peptide synthetases*.

These complexes are often laid out in a similar fashion, and they can contain many different modules to perform a diverse set of chemical manipulations on the developing product. These peptides are often cyclic and can have highly complex cyclic structures, although linear nonribosomal peptides are also common. Since the system is closely related to the machinery for building fatty acids and polyketides, hybrid compounds are often found. The presence of oxazoles or thiazoles often indicates that the compound was synthesized in this fashion.

Peptones

> Peptones are derived from animal milk or meat digested by proteolysis. In addition to containing small peptides, the resulting material includes fats, metals, salts, vitamins and many other biological compounds. Peptones are used in nutrient media for growing bacteria and fungi.

Peptide fragments

> Peptide fragments refer to fragments of proteins that are used to identify or quantify the source protein. Often these are the products of enzymatic degradation performed in the laboratory on a controlled sample, but can also be forensic or paleontological samples that have been degraded by natural effects.

Peptide Synthesis

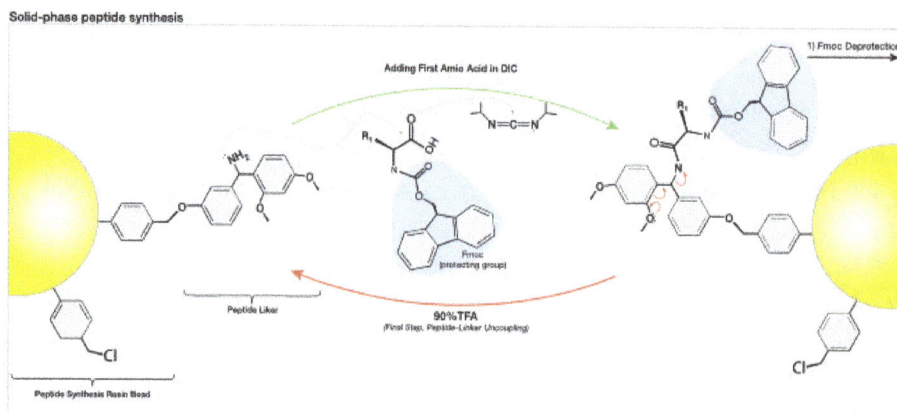

Solid-phase peptide synthesis on a rink amide resin using Fmoc-α-amine-protected amino acid

Peptides in Molecular Biology

Peptides received prominence in molecular biology for several reasons. The first is that peptides allow the creation of *peptide antibodies* in animals without the need of purifying the protein of interest. This involves synthesizing antigenic peptides of sections of the protein of interest. These will then be used to make antibodies in a rabbit or mouse against the protein.

Another reason is that peptides have become instrumental in mass spectrometry, allowing the identification of proteins of interest based on peptide masses and sequence.

Peptides have recently been used in the study of protein structure and function. For example, synthetic peptides can be used as probes to see where protein-peptide interactions occur

Inhibitory peptides are also used in clinical research to examine the effects of peptides on the inhibition of cancer proteins and other diseases. For example, one of the most promising application is through peptides that target LHRH. These particular peptides act as an agonist, meaning that they bind to a cell in a way that regulates LHRH receptors. The process of inhibiting the cell receptors suggests that peptides could be beneficial in treating prostate cancer. However, additional investigations and experiments are required before the cancer-fighting attributes, exhibited by peptides, can be considered definitive.

Well-known Peptide Families

The peptide families in this section are ribosomal peptides, usually with hormonal activity. All of these peptides are synthesized by cells as longer "propeptides" or "proproteins" and truncated prior to exiting the cell. They are released into the bloodstream where they perform their signaling functions.

Antimicrobial Peptides

- Magainin family
- Cecropin family

- Cathelicidin family
- Defensin family

Tachykinin Peptides

- Substance P
- Kassinin
- Neurokinin A
- Eledoisin
- Neurokinin B

Vasoactive Intestinal Peptides

- VIP (*Vasoactive Intestinal Peptide*; PHM27)
- PACAP *Pituitary Adenylate Cyclase Activating Peptide*
- Peptide PHI 27 (*Peptide Histidine Isoleucine 27*)
- GHRH 1-24 (*Growth Hormone Releasing Hormone 1-24*)
- Glucagon
- Secretin

Pancreatic Polypeptide-related Peptides

- NPY (*NeuroPeptide Y*)
- PYY (*Peptide YY*)
- APP (*Avian Pancreatic Polypeptide*)
- PPY *Pancreatic PolYpeptide*

Opioid Peptides

- Proopiomelanocortin (POMC) peptides
- Enkephalin pentapeptides
- Prodynorphin peptides

Calcitonin Peptides

- Calcitonin
- Amylin
- AGG01

Other Peptides

- B-type Natriuretic Peptide (BNP) - produced in myocardium & useful in medical diagnosis

- Lactotripeptides - Lactotripeptides might reduce blood pressure, although the evidence is mixed.

Notes on Terminology

Length:

- A *polypeptide* is a single linear chain of many amino acids, held together by amide bonds.

- A *protein* consists of one or more polypeptides (more than about 50 amino acids long).

- An *oligopeptide* consists of only a few amino acids (between two and twenty).

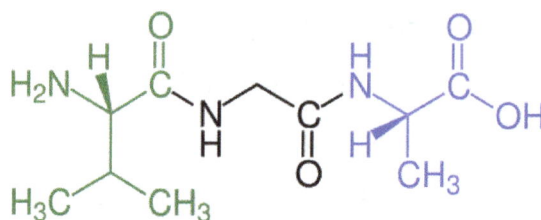

A tripeptide (example Val-Gly-Ala) with green marked amino end (L-Valine) and blue marked carboxyl end (L-Alanine)

Number of amino acids:

- A *monopeptide* has one amino acid.

- A *dipeptide* has two amino acids.

- A *tripeptide* has three amino acids.

- A *tetrapeptide* has four amino acids.

- A *pentapeptide* has five amino acids.

- A *hexapeptide* has six amino acids.

- A *heptapeptide* has seven amino acids.

- An *octapeptide* has eight amino acids (e.g., angiotensin II).

- A *nonapeptide* has nine amino acids (e.g., oxytocin).

- A *decapeptide* has ten amino acids (e.g., gonadotropin-releasing hormone & angiotensin I).

- An *undecapeptide* (or *monodecapeptide*) has eleven amino acids, a *dodecapeptide* (or *didecapeptide*) has twelve amino acids, a *tridecapeptide* has thirteen amino acids, and so forth.

- An *icosapeptide* has twenty amino acids, a *tricontapeptide* has thirty amino acids, a *tetracontapeptide* has forty amino acids, and so forth.

Function:

- A *neuropeptide* is a peptide that is active in association with neural tissue.

- A *lipopeptide* is a peptide that has a lipid connected to it, and *pepducins* are lipopeptides that interact with GPCRs.

- A *peptide hormone* is a peptide that acts as a hormone.

- A proteose is a mixture of peptides produced by the hydrolysis of proteins. The term is somewhat archaic.

Doping in Sports

The term *peptide* has been used to mean *secretagogue peptides* and *peptide hormones* in sports doping matters: secretagogue peptides are classified as Schedule 2 (S2) prohibited substances on the World Anti-Doping Agency (WADA) Prohibited List, and are therefore prohibited for use by professional athletes both in and out of competition. Such secretagogue peptides have been on the WADA prohibited substances list since at least 2008. The Australian Crime Commission cited the alleged misuse of secretagogue peptides in Australian sport including growth hormone releasing peptides CJC-1295, GHRP-6, and GHSR (gene) hexarelin. There is ongoing controversy on the legality of using secretagogue peptides in sports.

Peptide Bond

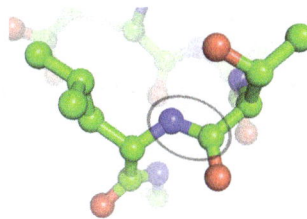

Peptide bond

A peptide bond (amide bond) is a covalent chemical bond linking two consecutive amino acid monomers along a peptide or protein chain.

Synthesis

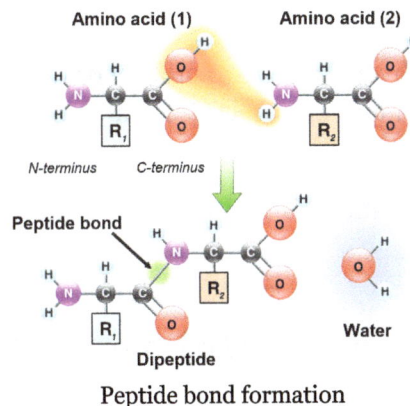

Peptide bond formation

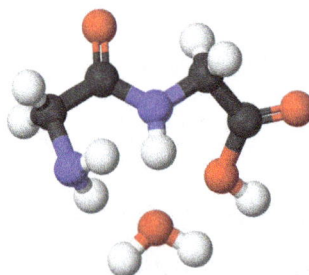

Glycine condensation

When two amino acids form a *dipeptide* through a *peptide bond* it is called condensation. In condensation, two amino acids approach each other, with the acid moiety of one coming near the amino moiety of the other. One loses a hydrogen and oxygen from its carboxyl group (COOH) and the other loses a hydrogen from its amino group (NH_2). This reaction produces a molecule of water (H_2O) and two amino acids joined by a peptide bond (-CO-NH-). The two joined amino acids are called a dipeptide.

The peptide bond is synthesized when the carboxyl group of one amino acid molecule reacts with the amino group of the other amino acid molecule, causing the release of a molecule of water (H_2O), hence the process is a dehydration synthesis reaction (also known as a condensation reaction).

The formation of the peptide bond consumes energy, which, in living systems, is derived from ATP. Polypeptides and proteins are chains of amino acids held together by peptide bonds. Living organisms employ enzymes to produce polypeptides, and ribosomes to produce proteins. Peptides are synthesized by specific enzymes. For example, the tripeptide glutathione is synthesized in two steps from free amino acids, by two enzymes: gamma-glutamylcysteine synthetase and glutathione synthetase.

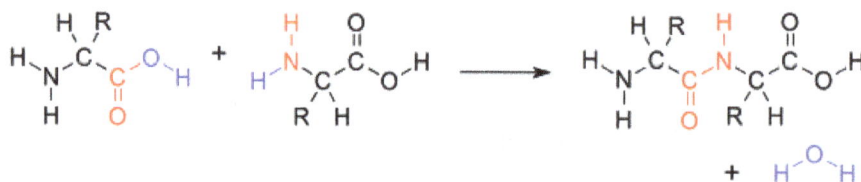

The condensation of two amino acids to form a peptide bond (red) with expulsion of water (blue)

Degradation

A peptide bond can be broken by hydrolysis (the addition of water). In the presence of water they will break down and release 8–16 kilojoule/mol (2–4 kcal/mol) of free energy. This process is extremely slow.

In living organisms, the process is catalyzed by enzymes known as peptidases or proteases.

Spectra

The wavelength of absorption for a peptide bond is 190–230 nm (which makes it particularly susceptible to UV radiation).

Dehydration synthesis (condensation) reaction forming an amide

Cis/Trans Isomers of the Peptide Group

Significant delocalisation of the lone pair of electrons on the nitrogen atom gives the group a partial double bond character. The partial double bond renders the amide group planar, occurring in either the cis or trans isomers. In the unfolded state of proteins, the peptide groups are free to isomerize and adopt both isomers; however, in the folded state, only a single isomer is adopted at each position (with rare exceptions). The trans form is preferred overwhelmingly in most peptide bonds (roughly 1000:1 ratio in trans:cis populations). However, X-Pro peptide groups tend to have a roughly 3:1 ratio, presumably because the symmetry between the C^α and C^δ atoms of proline makes the cis and trans isomers nearly equal in energy.

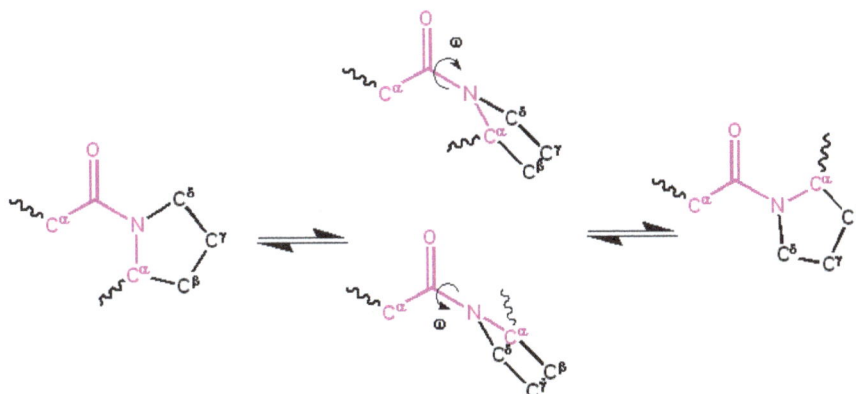

Isomerization of an X-Pro peptide bond. Cis and trans isomers are at far left and far right, respectively, separated by the transition states.

The dihedral angle associated with the peptide group (defined by the four atoms $C^\alpha - C' - N - C^\alpha$) is denoted ω; $\omega = 0°$ for the cis isomer (synperiplanar conformation) and $\omega = 180°$ for the trans isomer (antiperiplanar conformation). Amide groups can isomerize about the C'-N bond between the cis and trans forms, albeit slowly ($\tau \sim 20$ seconds at room temperature). The transition states $\omega = \pm 90°$ requires that the partial double bond be broken, so that the activation energy is roughly 80 kilojoule/mol (20 kcal/mol). However, the activation energy can be lowered (and the isomerization catalyzed) by changes that favor the single-bonded form, such as placing the peptide group in a hydrophobic environment or donating a hydrogen bond to the nitrogen atom of an X-Pro peptide group. Both of these mechanisms for lowering the activation energy have been observed in peptidyl prolyl isomerases (PPIases), which are naturally occurring enzymes that catalyze the cis-trans isomerization of X-Pro peptide bonds.

Conformational protein folding is usually much faster (typically 10–100 ms) than cis-trans isomerization (10–100 s). A nonnative isomer of some peptide groups can disrupt the conformational folding significantly, either slowing it or preventing it from even occurring until the native isomer is reached. However, not all peptide groups have the same effect on folding; nonnative isomers of other peptide groups may not affect folding at all.

Chemical Reactions

Due to its resonance stabilization, the peptide bond is relatively unreactive under physiological conditions, even less than similar compounds such as esters. Nevertheless, peptide bonds can undergo chemical reactions, usually through an attack of an electronegative atom on the carbonyl carbon, breaking the carbonyl double bond and forming a tetrahedral intermediate. This is the pathway followed in proteolysis and, more generally, in N-O acyl exchange reactions such as those of inteins. When the functional group attacking the peptide bond is a thiol, hydroxyl or amine, the resulting molecule may be called a cyclol or, more specifically, a thiacyclol, an oxacyclol or an azacyclol, respectively.

Chemistry of Peptide Bonds

- "Peptides" are small condensation products of amino acids

- They are "small" compared to proteins (di, tri, tetra... oligo-).

- C=N double bond character due to the resonance structure

- Restricted rotations about C=N resists hydrolysis

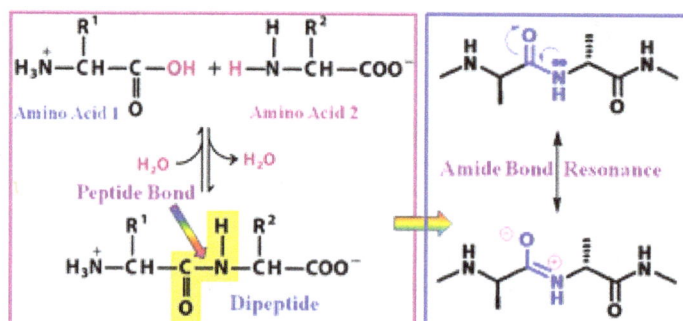

Reading the AA in a Peptide Chain

- By convention, peptide sequences are written left to right from the N-terminus to the C-terminus.

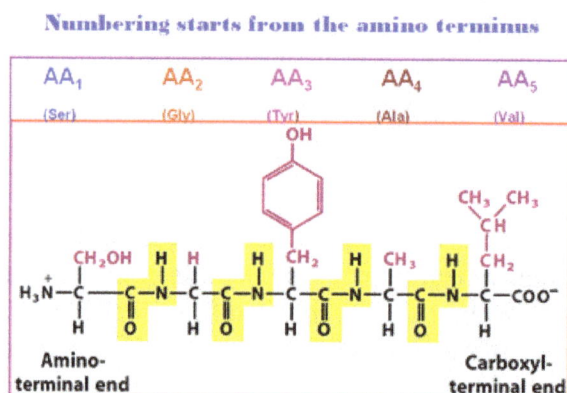

Naming a Peptide Sequence

1. Naming starts from the N-terminus

2. Sequence is written as: Ala-Glu-Gly-Lys

3. Sometimes the one-letter code is used: AEGK

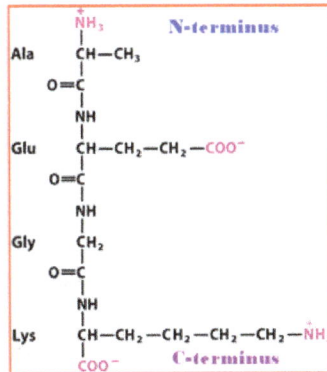

Peptides: A Variety of Functions

• Hormones and pheromones
  ➤ insulin (think sugar)
  ➤ oxytocin (think childbirth)
  ➤ sex-peptide (think fruit fly mating)
	• Neuropeptides
	  ➤ substance P (pain mediator)

• Antibiotics
  ➤ polymyxin B [for Gram (-) bacteria]
  ➤ bacitracin [for Gram (+) bacteria]
	• Protection, e.g. toxins
	  ➤ amanitin (mushrooms)
	  ➤ conotoxin (cone snails)
	  ➤ chlorotoxin (scorpions)

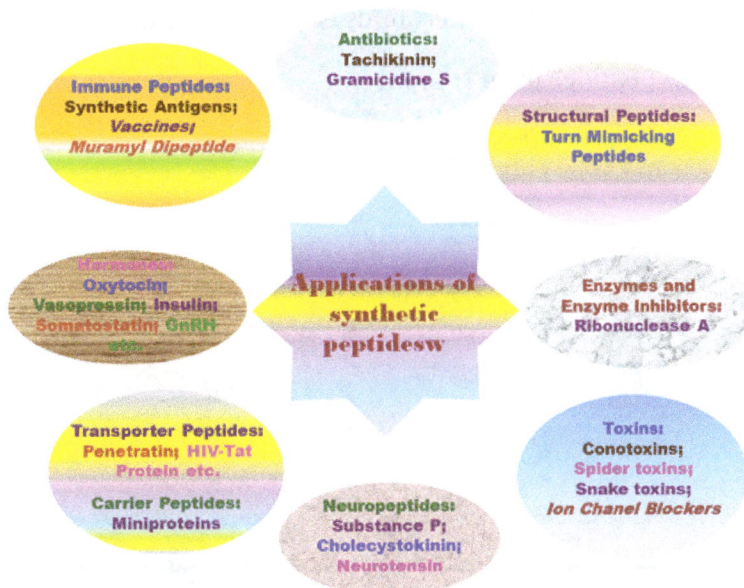

Schematic presentation to show a variety of applications of peptides.

Peptide Coupling: Need for Protecting Groups

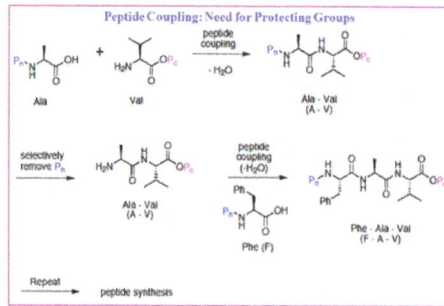

The Protecting Groups

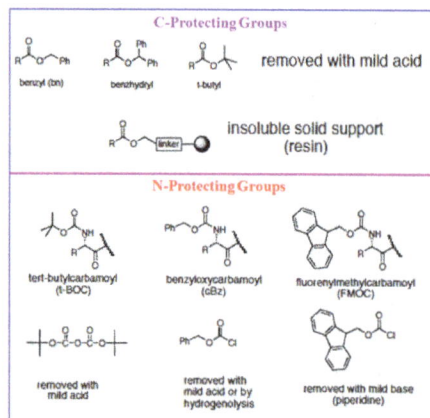

Solid-Phase Peptide Synthesis (SPPS)

- Peptides up to ~ 100 amino acids can be synthesized in a laboratory

- Laboratory synthesis is from the C-terminus to the N-terminus

- Nature synthesizes peptides from N to C.

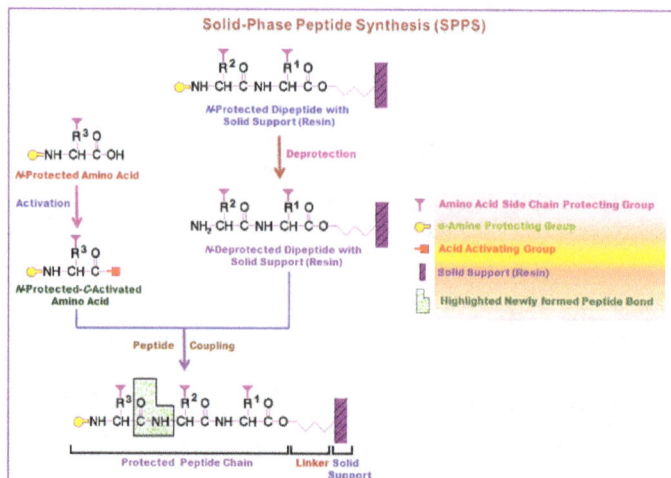

Solid phase peptide synthesis protocol

Solid-Phase Peptide Synthesis: The Solid Support

- Can be functionalised;

- Chemical stability (it must be inert to all applied chemicals);

- Mechanical stability (it shouldn't brake under stirring);

- It must swell extensively in the solvents used for the synthesis;

- Peptide-resin bond should be stable during the synthesis;

- Peptide-resin bond can be cleaved effectively at the end of the synthesis;

- The basic of the most common used resins is polystyrene-1,4-divinylbenzene (1-2%) copolymer.

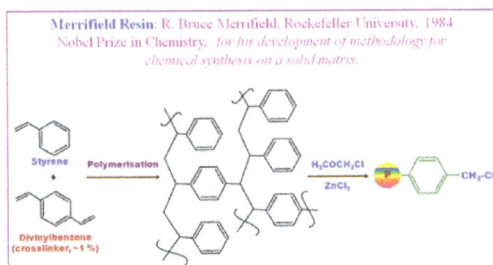

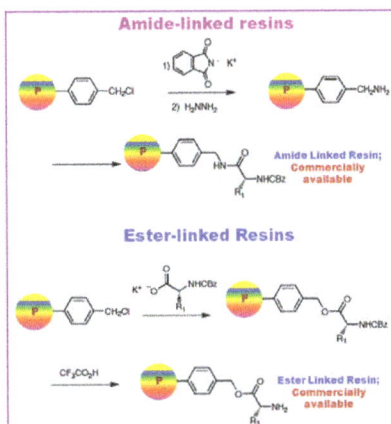

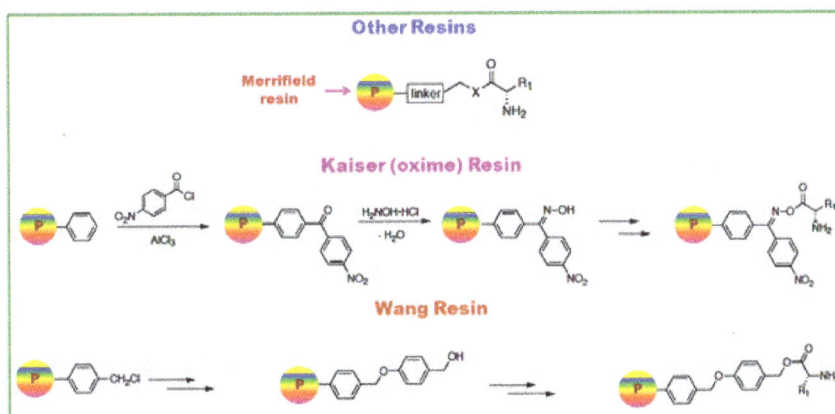

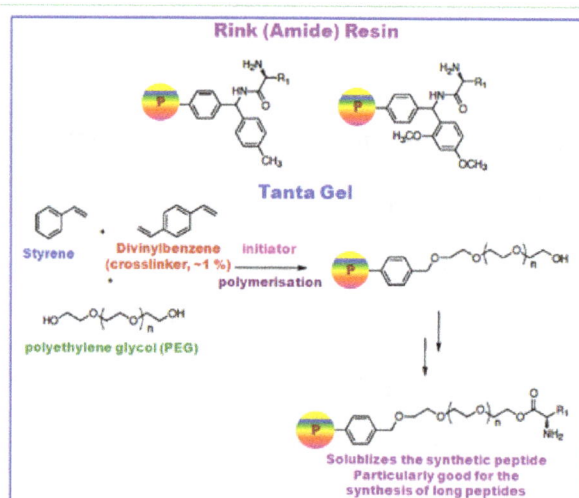

Peptide Coupling Reagent:

- N-protected carboxylic acid,

- C-protected amine

- DCC, HOBT

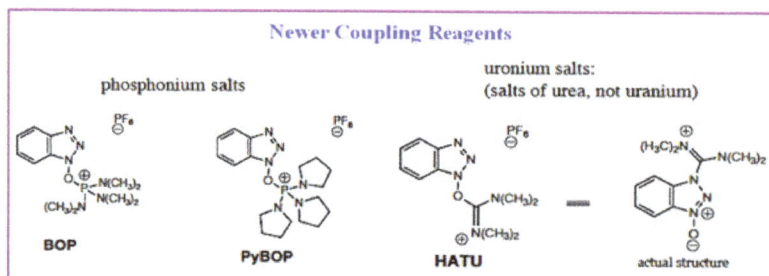

Why not N to C peptide synthesis?

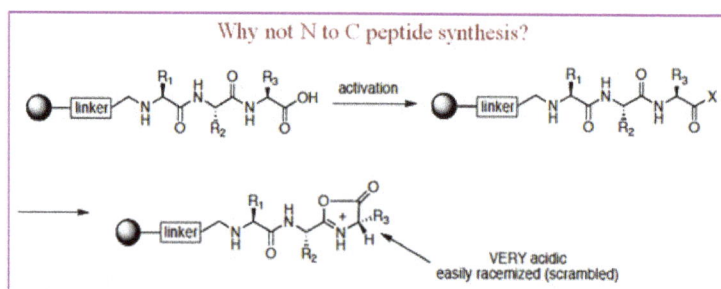

Importance of Maintaining Stereochemical Integrity During the Coupling Step

- Number of possible stereoisomers = 2n where n= no. of chiral centers

- A peptide w/ 10 AA residues has 210 possible stereoisomers.

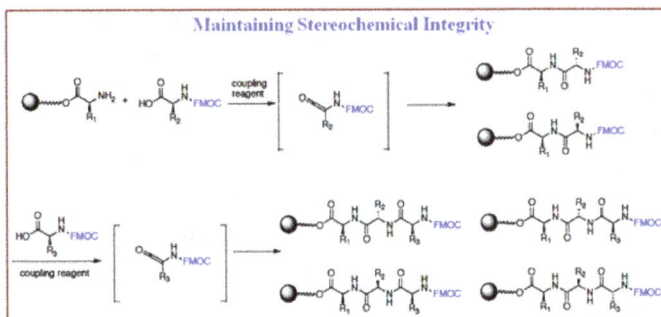

Proteins are:-

Natural β-amino Acids and β-peptides

β-peptides consist of β amino acids, which have their amino group bonded to the β- carbon rather than the α-carbon as in the 20 standard biological amino acids. The only commonly naturally occurring β-amino acid is β-alanine which is used as a component of larger bioactive molecules. However, β-peptides in general do not appear in nature. For this reason β-peptide-based antibiotics are being explored as ways of evading antibiotic resistance. Studies in this field were explored first in 1996 by the group of Dieter Seebach and that of Samuel Gellman.

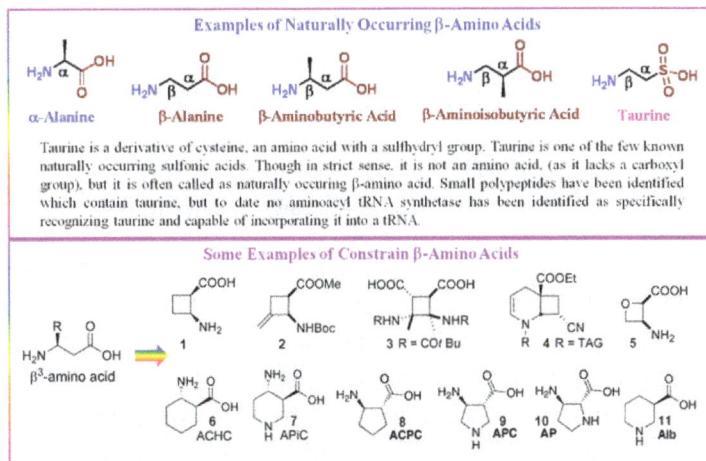

The chemical structures of naturally occurring and few unnatural β- amino acids.

The folded structures in natural polypeptides containing α-amino acids are conveniently defined using backbone torsion angles Φ and ψ. Each α-amino acid residue possesses two degrees of torsional freedom about the N–Cα (Φ) and Cα-CO (ψ) bonds, with the peptide bond restricted to a trans, planar (ω ≈ 180°) conformation. The Ramachandran map provides a convenient means of analyzing and representing the observed backbone conformations of α-amino acid residues in peptides and proteins. A very large numbers of natural and unnatural amino acids have been utilized to generate mimic many biologically and structurally interesting compounds peptides and proteins. Gellman and co-workers have studied several model peptides containing covalently constrained β-amino acids. Their study led to the realization that new classes of folded unnatural polypeptide structures could be generated. The insertion of additional atoms in between the flanking peptide units enhances the number of degrees of torsional freedom, resulting in an expansion of energetically accessible conformational space. As for example, in the β-amino acid residue, local backbone conformations are determined by values of three torsional variables (Φ, θ, and ψ) while for the γ-residue, the number of torsional variables is four (Φ, θ1, θ2, and ψ).

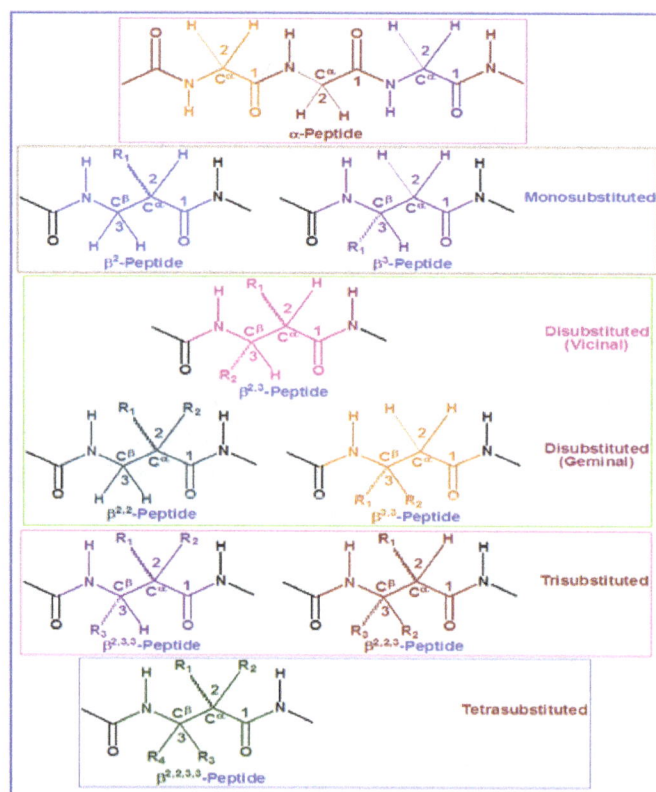

Substitution patterns for a β-amino acid residue.

A poly-β-amino acid helix introduced by Kovac et al. in 1965 has 3.4 residues/turn and an axial translation of 1.58 Å/ residue. It is necessarily right-handed for the β-amino acids with the L- configuration at the α-carbon. Recent interest in the chemistry and biology of peptides containing backbone expanded amino acid residues stems from the studies reported in the mid-1990s that novel polypeptide helices could be formed in oligo β-peptides and the characterization of hybrid structures that demonstrated that the β-and γ-residues can be accommodated into canonical helical folds, with an expansion of the intramolecular hydrogen bonded rings.

Peptide Synthesis

In organic chemistry, peptide synthesis is the production of peptides, which are organic compounds in which multiple amino acids are linked via amide bonds, also known as peptide bonds. The biological process of producing long peptides (proteins) is known as protein biosynthesis.

Chemistry

Peptides are synthesized by coupling the carboxyl group of one amino acid to the amino group of another amino acid molecule. Due to the possibility of unintended reactions, protecting groups are usually necessary. Chemical peptide synthesis most commonly starts at the carboxyl end of the peptide, and proceeds toward the amino-terminus. This is the opposite direction of protein biosynthesis.

Liquid-phase Synthesis

Liquid-phase peptide synthesis is a classical approach to peptide synthesis. It has been replaced in most labs by solid-phase synthesis. However, it retains usefulness in large-scale production of peptides for industrial purposes.

Solid-phase Synthesis

Solid-phase peptide synthesis (SPPS), pioneered by Robert Bruce Merrifield, caused a paradigm shift within the peptide synthesis community, and it is now the standard method for synthesizing peptides and proteins in the lab. SPPS allows for the synthesis of natural peptides which are difficult to express in bacteria, the incorporation of unnatural amino acids, peptide/protein backbone modification, and the synthesis of D-proteins, which consist of D-amino acids.

Small porous beads are treated with functional units ('linkers') on which peptide chains can be built. The peptide will remain covalently attached to the bead until cleaved from it by a reagent such as anhydrous hydrogen fluoride or trifluoroacetic acid. The peptide is thus 'immobilized' on the solid-phase and can be retained during a filtration process while liquid-phase reagents and by-products of synthesis are flushed away.

The general principle of SPPS is one of repeated cycles of deprotection-wash-coupling-wash. The free N-terminal amine of a solid-phase attached peptide is coupled to a single N-protected amino acid unit. This unit is then deprotected, revealing a new N-terminal amine to which a further amino acid may be attached. The superiority of this technique partially lies in the ability to perform wash cycles after each reaction, removing excess reagent with all of the growing peptide of interest remaining covalently attached to the insoluble resin.

The overwhelmingly important consideration is to generate extremely high yield in each step. For example, if each coupling step were to have 99% yield, a 26-amino acid peptide would be synthesized in 77% final yield (assuming 100% yield in each deprotection); if each step were 95%, it would be synthesized in 25% yield. Thus each amino acid is added in major excess (2~10x) and coupling amino acids together is highly optimized by a series of well-characterized agents.

There are two majorly used forms of SPPS – *Fmoc* and *Boc*. Unlike ribosome protein synthesis, solid-phase peptide synthesis proceeds in a C-terminal to N-terminal fashion. The N-termini of amino acid monomers is protected by either of these two groups and added onto a deprotected amino acid chain. Automated synthesizers are available for both techniques, though many research groups continue to perform SPPS manually.

SPPS is limited by yields, and typically peptides and proteins in the range of 70 amino acids are pushing the limits of synthetic accessibility. Synthetic difficulty also is sequence dependent; typically amyloid peptides and proteins are difficult to make. Longer lengths can be accessed by using native chemical ligation to couple two peptides together with quantitative yields.

Since its introduction over 40 years ago, SPPS has been significantly optimized. First, the resins themselves have been optimized. Furthermore, the 'linkers' between the C-terminal amino acid and polystyrene resin have improved attachment and cleavage to the point of mostly quantitative yields. The evolution of side chain protecting groups has limited the frequency of unwanted side reactions. In addition, the evolution of new activating groups on the carboxyl group of the incoming amino acid have improved coupling and decreased epimerization. Finally, the process itself has been optimized. In Merrifield's initial report, the deprotection of the α-amino group resulted in the formation of a peptide-resin salt, which required neutralization with base prior to coupling. The time between neutralization of the amino group and coupling of the next amino acid allowed for aggregation of peptides, primarily through the formation of secondary structures, and adversely affected coupling. The Kent group reported that concomitant neutralization of the α-amino group and coupling of the next amino acid led to improved coupling. Each of these improvements has helped SPPS become the robust technique that it is today.

Solid Supports

The name solid support implies that reactions are carried out on the surface of the support, but this is not the case. Reactions also occur within these particles, and thus the term "solid support" better describes the insolubility of the polymer. The physical properties of the solid support, and the applications to which it can be utilized, vary with the material from which the support is constructed, the amount of cross-linking, as well as the linker and handle being used. Most scientists in the field believe that supports should have the minimum amount of cross-linking to confer stability. This should result in a well-solvated system where solid-phase peptide synthesis can be carried out. Nonetheless, the characteristics of an efficient solid support include: it must be physically stable and permit the rapid filtration of liquids, such as excess reagents; it must be inert to all reagents and solvents used during SPPS; it must swell extensively in the solvents used to allow for penetration of the reagents; and it must allow for the attachment of the first amino acid.

There are three primary types of solid supports: gel-type supports, surface-type supports, and composites. Gel-type supports are highly solvated polymers with an equal distribution of functional groups. This type of support is the most common, and includes polystyrene (styrene cross-linked with 1–2% divinylbenzene), polyacrylamide (hydrophilic alternative to polystyrene), polyethylene glycol (PEG) (PEG-polystyrene (PEG-PS) is more stable than polystyrene and spaces the site of synthesis from the polymer backbone), and PEG-based supports composed of a PEG-polypropylene glycol network or PEG with polyamide or polystyrene. Surface-type supports: Many materials have been developed for surface functionalization, including controlled pore glass, cellulose fibers, and highly cross-linked polystyrene. Composites are gel-type polymers supported by rigid matrices.

Polystyrene Resin

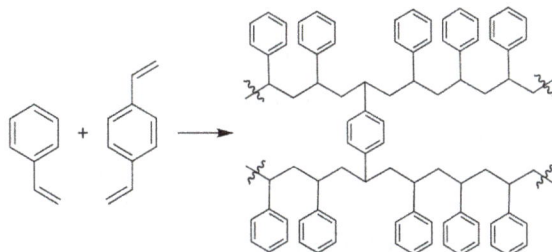

Polystyrene cross-linked with divinylbenzene. This is the most common solid support used in SPPS, and was the support pioneered by R. Bruce Merrifield.

Polystyrene resin is a versatile resin and it is quite useful in multi-well, automated peptide synthesis, due to its minimal swelling in dichloromethane. The initial support used by R. Bruce Merrifield was polysytrene cross-linked with 2% divinylbenzene. This support is sometimes referred to as the 'Merrifield resin.' This resin produces a hydrophobic bead that is solvated by a nonpolar solvent such as dichloromethane or dimethylformamide. Since then, new resins have been developed that have the advantages of chemical inertness, and enhanced swelling or rigidity (a property of mechanical strength). For instance, highly cross-linked (50%) polystyrene has been developed that possesses the features of increased mechanical stability, better filtration of reagents and solvents, and rapid reaction kinetics.

Polystyrene resins are also available as PEG hybrids. An example of this type of resin is the Tentagel resin. The base resin is polystyrene onto which is attached long chains (MW ca. 3000 Da) of polyethylene glycol (PEG; also known as polyethylene oxide). Synthesis is carried out on the distal end of the PEG spacer making it suited for long and difficult peptides. In addition, it is also attractive for the synthesis of combinatorial Peptide libraries and on resin screening experiments. It does not expand much during synthesis, making it a preferred resin for robotic peptide synthesis.

Polyamide Resin

Polyamide resin is also a useful and versatile resin. It seems to swell much more than polystyrene, in which case it may not be suitable for some automated synthesizers, if the wells are too small.

PEG-based Resin

Improvements to solid supports used for peptide synthesis enhance their ability to withstand the repeated use of TFA during the deprotection step of SPPS. Furthermore, different resins allow for different functional groups at the C-terminus. The oxymethylphenylacetamidomethyl (PAM) resin results in the conventional C-terminal carboxylic acid. On the other hand, the paramethylbenzhydrylamine (pMBHA) resin yields a C-terminal amide, which is useful in mimicking the interior of a protein.

Along with the development of Fmoc SPPS, different resins have also been created to be removed by TFA. Similar to the Boc strategy, two primary resins are used, based on whether a C-terminal carboxylic acid or amide is desired. The Wang resin was, as of 1996, the most commonly used resin for peptides with C-terminal carboxylic acids. If a C-terminal amide is desired, the Rink amide resin is used.

Protecting Groups

Amino acids have reactive α-carboxylic acid and α-amine groups that allow for their linking into polymers, but that also complicate the aim coupling specific pairs of amino acids, in precise order. In addition, many amino acids have reactive side chain functional groups, which can also react in a variety of ways, including with free α-carboxylic and α-amino groups during peptide synthesis (given the very reactive reagents present), "side-reactions" that would negatively influence yield and purity. Chemical groups have been developed to facilitate the synthesis of peptides with precise amino acid sequences, with minimal side reactions, groups that block or "protect" all functional groups present in amino acids except that pair whose coupling is desired. These protecting groups, while very many in practice, can be described in three groups: α-carboxylic acid (C-terminal) protecting groups, α-amino (N-terminal) protecting groups, and side chain protecting groups.

C-terminal Protecting Groups

Protecting groups of the carboxylic acid are the least used given the direction and methods applied in solid-supported peptide synthesis, but are covered first because to the brevity of the topic. These protecting groups are mostly used in liquid-phase synthesis, and there is some redundancy in the chemistry with respect to groups used to protect carboxylates in side chains.

N-terminal Protecting Groups

Protected amino acids being added in a given step of peptide synthesis are added in excess to ensure maximal yields during each synthesis step. Without N-terminal protection of the added amino acid, its self-coupling (polymerization) would compete with the desired peptide synthesis, resulting in low yield of even failure of the desired peptide synthesis. N-terminal protection requires an additional step of removing the protecting group in each peptide synthesis cycle, a deprotection step, prior to the next coupling step.

As of this date, two protecting groups, tert-butyloxycarbonyl (t-Boc) and 9H-fluoren-9-ylmethoxycarbonyl (Fmoc) are most commonly used to protect the α-amine group of the "incoming" (newly added) amino acid in a solid-phase peptide synthesis cycle.

t-Boc and Fmoc Protecting Groups

Tert-butyloxycarbonyl (t-Boc) Protection

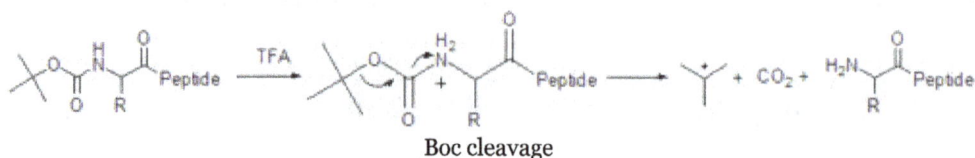

Boc cleavage

The original method for the synthesis of proteins relied on tert-butyloxycarbonyl (or more simply "Boc") to temporarily protect the α-amino group. In this method, the Boc group is covalently bound to the amino group to suppress its nucleophilicity. The C-terminal amino acid is covalently linked to the resin through a linker. Next, the Boc group is removed with acid, such as trifluoroacetic acid (TFA). This forms a positively charged amino group (in the presence of excess TFA; note image on the right illustrates neutral amino group), which is neutralized (via in-situ or non-in-situ methods) and coupled to the incoming activated amino acid. Reactions are driven to completion by the use of excess (2- to 4-fold) activated amino acid. After each deprotection and coupling step, a wash with dimethylformamide (DMF) is performed to remove excess reagents, allowing (by year 2000) for high yields (~99%) during each cycle.

t-Boc protecting strategies retain usefulness in reducing peptide aggregation during synthesis. *t*-Boc groups can be added to amino acids with *t*-Boc anhydride and a suitable base. Some researchers prefer Boc SPPS for complex syntheses . In addition, when synthesizing nonnatural peptide analogs, which are base-sensitive (such as depsipeptides), the *t*-Boc protecting group is necessary, because Fmoc SPPS uses a base to deprotect the α-amino group.

Permanent side-chain protecting groups are typically benzyl or benzyl-based groups. Final removal of the peptide from the linkage occurs simultaneously with side-chain deprotection with anhydrous hydrogen fluoride via hydrolytic cleavage. The final product is a fluoride salt which is relatively easy to solubilize. Importantly, scavengers such as cresol are added to the HF in order to prevent reactive *t*-butyl cations from generating undesired products. In fact, the use of harsh hydrogen fluoride may degrade some peptides, which was the premise for the development of a milder, base-labile method of SPPS—namely, the Fmoc method.

9H-fluoren-9-ylmethoxycarbonyl (Fmoc) Protection

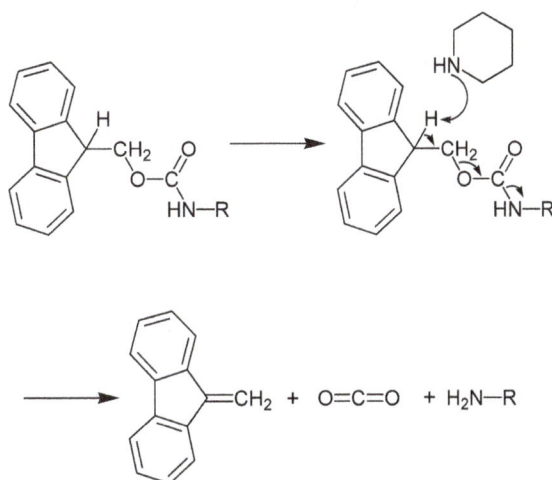

Cleavage of the amine-protecting Fmoc group.

Treatment of the Fmoc-protected amino acid, abbreviated by "R," with piperidine, resulting in proton abstraction from the methine group of the fluorenyl ring system, which results in cleavage of that ring system as dibenzofulvene, and release of the amino acid as a carbamate, which decomposes to carbon dioxide (CO_2) and the amino acid moiety with free amine group. The gentleness of the deprotection is believed to lie in the acidity of the fluorenyl proton, resulting from stabilization of the resulting aromatic anion. Absent a scavenger (e.g., such as a thiol), the dibenzofulvene product can react with nucleophiles such as the piperidine (which is in large excess), or even the released amine.

The capacity for anhydrous hydrogen fluoride to degrade proteins during the final cleavage conditions led to a new α-amino protecting group based on 9-fluorenylmethyloxycarbonyl (Fmoc). The Fmoc method allows for a milder deprotection scheme. This method utilizes a base, usually piperidine (20–50%) in DMF in order to remove the Fmoc group to expose the α-amino group for reaction with an incoming activated amino acid. Unlike the acid used to deprotect the α-amino group in Boc methods, Fmoc SPPS uses a base, and thus the exposed amine is neutral. Therefore, no neutralization of the peptide-resin is required, but the lack of electrostatic repulsions between the peptides can lead to increased aggregation. Because the liberated fluorenyl group is a chromophore, deprotection by Fmoc can be monitored by UV absorbance of the runoff, a strategy which is employed in automated synthesizers.

The advantage of Fmoc is that it is cleaved under very mild basic conditions (e.g. piperidine), but stable under acidic conditions, although this has not always held true in certain synthetic sequences. This allows mild acid-labile protecting groups that are stable under basic conditions, such as Boc and benzyl groups, to be used on the side-chains of amino acid residues of the target peptide. This orthogonal protecting group strategy is common in organic synthesis. Fmoc is preferred over BOC due to ease of cleavage; however it is less atom-economical, as the fluorenyl group is much larger than the tert-butyl group. Accordingly, prices for Fmoc amino acids were high until the large-scale piloting of one of the first synthesized peptide drugs, enfuvirtide, began in the 1990s, when market demand adjusted the relative prices of the two sets of amino acids.

Semipermanent side chain protecting groups are t-butyl-based, and final cleavage of the protein from the resin and removal of permanent protecting groups is performed with TFA in the presence of scavengers. Water and triisopropylsilane (TIPS) present in a 1:1 ratio are often used as scavengers. Thus, the Fmoc method is orthogonal in two directions: deprotection of any α-amino group, deprotection of side groups and final cleavage from the resin occur by independent mechanisms. The resulting final product is a TFA salt, which is more difficult to solubilize than the fluoride salts generated in Boc SPPS. This method is thus milder than the Boc method because the deprotection/cleavage-from-resin steps occur with different conditions rather than with different reaction rates.

Comparison of *t*-Boc and Fmoc Solid-phase Peptide Synthesis

Both the Fmoc and Boc methods offer advantages and disadvantages. The selection of one technique over another is thus made on a case-by-case basis.

	Boc	Fmoc
Requires special equipment	Yes	No
Cost of reagents	Lower	Higher
Solubility of peptides	Higher	Lower
Purity of hydrophobic peptides	High	May be lower
Problems with aggregation	Less frequently	More frequently
Synthesis time	~20 min/amino acid	~20–60 min/amino acid
Cleavage from resin	HF	TFA
Safety	Potentially dangerous	Relatively safe
Orthogonal	No	Yes

Boc SPPS uses special equipment to handle the final cleavage and deprotection step, which requires anhydrous hydrogen fluoride. Because the final cleavage of the peptide with Fmoc SPPS uses TFA, this special equipment is not necessary. The solubility of peptides generated by Boc SPPS is generally higher than those generated with the Fmoc method, because fluoride salts are higher in solubility than TFA salts. Next, problems with aggregation are generally more of an issue with Fmoc SPPS, primarily because the removal of a Boc group with TFA yields a positively charged α-amino group, whereas the removal of an Fmoc group yields a neutral α-amino group. The electrostatic repulsion of the positively charged α-amino group limits the formation of secondary structure on the resin. Finally, the Fmoc method is considered orthogonal, since α-amino group deprotection is with base, while final cleavage from the resin is with acid. The Boc method utilizes acid for both deprotection and cleavage from the resin. Hence, both methods possess advantages and disadvantages for their application in specific situations, and several factors must be considered to decide between the methods.

Other Protecting Groups

Benzyloxy-carbonyl

The (Z) group is another carbamate-type amine protecting group, first used by Max Bergmann in the synthesis of oligopeptides. It is removed under harsh conditions using HBr in acetic acid, or milder conditions of catalytic hydrogenation. While it has been used periodically for α-amine protection in peptide synthesis, of this date, it is almost exclusively used for side chain protection.

Alloc and Miscellaneous Groups

The allyloxycarbonyl (alloc) protecting group is sometimes used to protect an amino group (or carboxylic acid or alcohol group) when an orthogonal deprotection scheme is required. It is also sometimes used when conducting on-resin cyclic peptide formation, where the peptide is linked to the resin by a side-chain functional group. The Alloc group can be removed using tetrakis(triphenylphosphine)palladium(0).

For special applications like synthetic steps involving protein microarrays, protecting groups sometimes termed "lithographic" are used, which are amenable to photochemistry at a particular wavelength of light, and so which can be removed during lithographic types of operations.

Side Chain Protecting Groups

Amino acid side chains represent a broad range of functional groups and are sites of nonspecific reactivity during peptide synthesis. Because of this, many different protecting groups are required that are usually based on the benzyl (Bzl) or tert-butyl (tBu) group. The specific protecting groups used during the synthesis of a given peptide vary depending on the peptide sequence and the type of N-terminal protection used. Side chain protecting groups are known as permanent or semipermanent protecting groups, because they can withstand the multiple cycles of chemical treatment during synthesis and are only removed during treatment with strong acids after peptide synthesis is completed.

Because N-terminal deprotection occurs repeatedly during peptide synthesis, protecting schemes are established such that different types of side chain protecting groups (e.g., Bzl, tBu, etc.) are matched to either Boc or Fmoc, respectively, for ultimate deprotection that is optimized. Because multiple protecting groups are normally used during peptide synthesis of peptides greater in length than oligomers, care must be taken such that all side chain protecting groups are compatible, so that when deprotection of individual protecting groups is required (e.g., to selectively modify the side chain of one amino acid of a synthetic peptide), the one deprotection step does not affect other side chains. Protecting schemes are therefore developed in each particular case of a peptide synthesis to assign protecting groups to each amino acid residue.

Activating Groups

For coupling the peptides the carboxyl group is usually activated. This is important for speeding up the reaction. There are two main types of activating groups: carbodiimides and triazolols. However the use of pentafluorophenyl esters (FDPP, PFPOH) and BOP-Cl are useful for cyclising peptides.

Carbodiimides

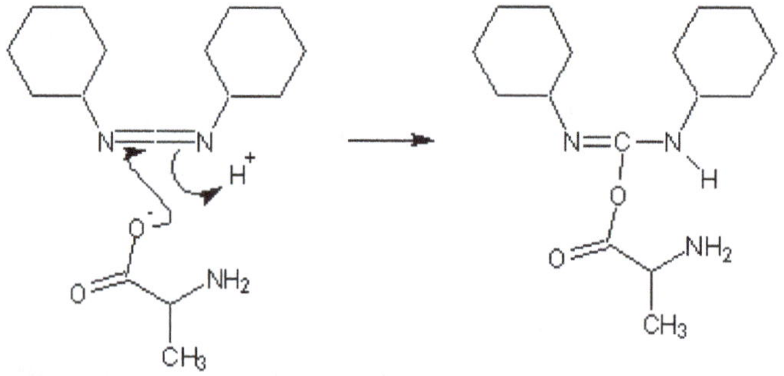

Alanine attaching to DCC

These activating agents were first developed. Most common are dicyclohexylcarbodiimide (DCC) and diisopropylcarbodiimide (DIC). Reaction with a carboxylic acid yields a highly reactive O-acyli-sourea. During artificial protein synthesis (such as Fmoc solid-state synthesizers), the C-terminus is often used as the attachment site on which the amino acid monomers are added. To enhance the electrophilicity of carboxylate group, the negatively charged oxygen must first be "activated" into a better leaving group. DCC is used for this purpose. The negatively charged oxygen will act as a nucleophile, attacking the central carbon in DCC. DCC is temporarily attached to the former carboxylate group (which is now an ester group), making nucleophilic attack by an amino group (on the attaching amino acid) to the former C-terminus (carbonyl group) more efficient. The problem with carbodiimides is that they are too reactive and that they can therefore cause racemization of the amino acid.

Triazoles

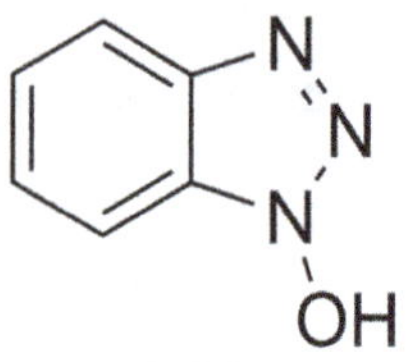

HOBt

HOAt

resin

Neighbouring group effect of HOAt

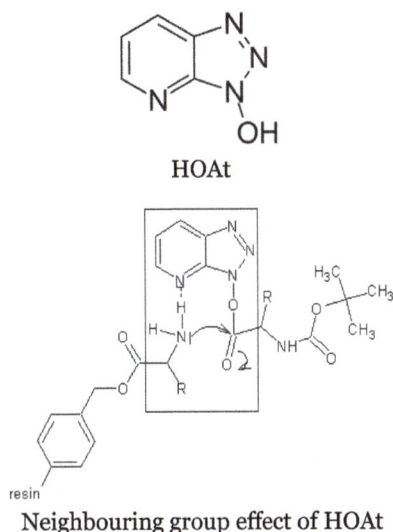

To solve the problem of racemization, triazoles were introduced. The most important ones are 1-hydroxy-benzotriazole (HOBt) and 1-hydroxy-7-aza-benzotriazole (HOAt), though others have also been developed. These substances can react with the O-acylurea to form an active ester which is less reactive and less in danger of racemization. HOAt is especially favourable because of a neighbouring group effect. As of this date, HOBt has been removed from many chemical vendor catalogues; although almost always found as a hydrate, HOBt may be explosive when allowed to fully dehydrate and shipment by air or sea is heavily restricted. Alternatives to HOBt and HOAt have been introduced. One of the most promising and inexpensive is ethyl 2-cyano-2-(hydroxyimino)acetate (trade name Oxyma Pure), which is not explosive and has a reactivity of that in between HOBt and HOAt.

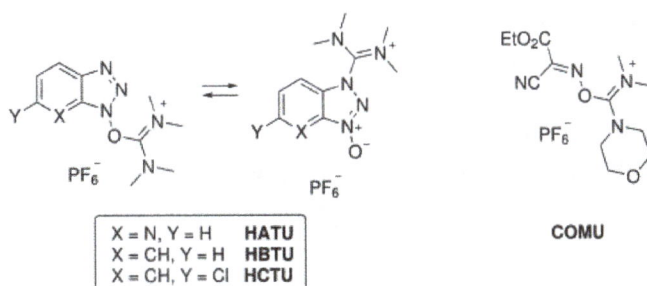

X = N, Y = H	HATU
X = CH, Y = H	HBTU
X = CH, Y = Cl	HCTU

COMU

Uronium-based peptide coupling reagents

Newer developments omit the carbodiimides totally: the active ester is introduced as a uronium or phosphonium salt of a non-nucleophilic anion (tetrafluoroborate or hexafluorophosphate): HBTU, HATU, HCTU, TBTU, PyBOP. Two uronium types of the coupling additive of Oxyma Pure is also available as COMU or TOTU reagent.

Regioselective Disulfide Bond Formation

The formation of multiple native disulfides remains one of the primary challenges of native peptide synthesis by solid-phase methods. Random chain combination typically results in several products with nonnative disulfide bonds. Stepwise formation of disulfide bonds is typically the preferred method, and performed with thiol protecting groups. Different thiol protecting groups provide multiple dimensions of orthogonal protection. These orthogonally protected cysteines are incorporated during the solid-phase synthesis of the peptide. Successive removal of these groups, to allow for selective exposure of free thiol groups, leads to disulfide formation in a stepwise manner. The order of removal of the groups must be considered so that only one group is removed at a time. Using this method, Kiso and coworkers reported the first total synthesis of insulin in 1993.

Thiol protecting groups used in peptide synthesis requiring later regioselective disulfide bond formation must possess multiple characteristics. First, they must be reversible with conditions that do not affect the unprotected side chains. Second, the protecting group must be able to withstand the conditions of solid-phase synthesis. Third, the removal of the thiol protecting group must be such that it leaves intact other thiol protecting groups, if orthogonal protection is desired. That is, the removal of PG A should not affect PG B. Some of the thiol protecting groups commonly used include the acetamidomethyl (Acm), tert-butyl (But), 3-nitro-2-pyridine sulfenyl (NPYS), 2-pyridine-sulfenyl (Pyr), and triphenylmethyl (Trt) groups. Importantly, the NPYS group can replace the Acm PG to yield an activated thiol.

Important to the discussion of disulfide bond formation is the order in which disulfides are formed. The synthesis insulin by Kiso and coworkers is illustrative of the logic and methods for regioselective disulfide bond formation. In this work, the A-chain of insulin was prepared with following protecting groups in place on its cysteines: CysA6(But), CysA7(Acm), and CysA11(But), leaving CysA20 unprotected. Synthesis of the B-chain was performed with the following protecting groups in place CysB7(Acm) and CysB19(Pyr). The first disulfide bond, CysA20–CysB19, was formed by mixing the two chains in 8 M urea, pH 8 (RT) for 50 min, while the second disulfide bond, CysA7–CysB7, was formed by treatment with iodine in aqueous acetic acid to remove the Acm groups. The third disulfide, the intramolecular CysA6–CysA11, was formed after the removal of the But groups by methyltrichlorosilane with diphenyl sulfoxide in TFA. Importantly, formation of the first disulfide in 8 M urea, pH 8 does not affect the other protecting groups, namely Acm and But groups. Likewise, formation of the second disulfide bond with iodine in aqueous acetic acid does not affect the But groups. From a logical standpoint, the order in which the thiol groups are exposed to form disulfides should be of little consequence, since the other cysteines are protected; however, it is observed, practically, that the order in which disulfides are formed can have a significant effect on yields.

Synthesizing Long Peptides

Stepwise elongation, in which the amino acids are connected step-by-step in turn, is ideal for small peptides containing between 2 and 100 amino acid residues. Another method is fragment condensation, in which peptide fragments are coupled. Although the former can elongate the peptide chain without racemization, the yield drops if only it is used in the creation of long or highly polar peptides. Fragment condensation is better than stepwise elongation for synthesizing sophisticated long peptides, but its use must be restricted in order to protect against racemization. Fragment condensation is also undesirable since the coupled fragment must be in gross excess, which may be a limitation depending on the length of the fragment.

A new development for producing longer peptide chains is chemical ligation: unprotected peptide chains react chemoselectively in aqueous solution. A first kinetically controlled product rearranges to form the amide bond. The most common form of native chemical ligation uses a peptide thioester that reacts with a terminal cysteine residue. Other methods applicable for covalently linking polypeptides in aqueous solution include the use of split inteins, spontaneous isopeptide bond formation and sortase ligation.

In order to optimize synthesis of long peptides, a method was developed in Medicon Valley for converting peptide sequences. The simple pre-sequence (e.g. Lysine (Lysn); Glutamic Acid (Glun); (LysGlu)n) that is incorporated at the C-terminus of the peptide to induce an alpha-helix-like structure. This can potentially increase biological half-life, improve peptide stability and inhibit enzymatic degradation without altering pharmacological activity or profile of action.

Microwave-assisted Peptide Synthesis

Although microwave irradiation has been around since the late 1940s, it was not until 1986 that microwave energy was used in organic chemistry. During the end of the 1980s and 1990s, microwave energy was an obvious source for completing chemical reactions in minutes that would otherwise take several hours to days. Through several technical improvements at the end of the 1990s and beginning of the 2000s, microwave synthesizers have been designed to provide both low and high energy pockets of microwave energy so that the temperature of the reaction mixture could be controlled. The microwave energy used in peptide synthesis is of a single frequency providing maximum penetration depth of the sample which is in contrast to conventional kitchen microwaves.

In peptide synthesis, microwave irradiation has been used to complete long peptide sequences with high degrees of yield and low degrees of racemization. Microwave irradiation during the coupling of amino acids to a growing polypeptide chain is catalyzed not only by the increase in temperature but also by the alternating electric field of the microwave. This is because the polar N-terminal amine group and peptide backbone continuously try to align with the alternating electric field, thus helping prevent aggregation and increasing access to the solid phase reaction matrix. This increases yields of the final peptide products. There is however no clear evidence that microwave is better than simple heating and some peptide laboratories regard microwave just as a convenient method for rapid heating of the peptidyl resin. Heating to above 50–55 degrees Celsius also prevents aggregation and accelerates the coupling.

Despite the main advantages of microwave irradiation of peptide synthesis, the main disadvantage is the racemization which may occur with the coupling of cysteine and histidine. A typical coupling reaction with these amino acids are performed at lower temperatures than the other 18 natural amino acids. A number of peptides do not survive microwave synthesis or heating in general. One of the more serious side effects is dehydration (loss of water) which for certain peptides can be almost quantitative like pancreatic polypeptide (PP). This side effect is also seen by simple heating without the use of microwave.

Cyclic Peptides

On Resin Cyclization

Peptides can be cyclized on a solid support. A variety of cylization reagents can be used such as HBTU/HOBt/DIEA, PyBop/DIEA, PyClock/DIEA. Head-to-tail peptides can be made on the solid support. The deprotection of the C-terminus at some suitable point allows on-resin cyclization by amide bond formation with the deprotected N-terminus. Once cyclization has taken place, the peptide is cleaved from resin by acidolysis and purified. The strategy for the solid-phase synthesis of cyclic peptides in not limited to attachment through Asp, Glu or Lys side chains. Cysteine has a very reactive sulfhydryl group on its side chain. A disulfide bridge is created when a sulfur atom from one Cysteine forms a single covalent bond with another sulfur atom from a second cysteine in a different part of the protein. These bridges help to stabilize proteins, especially those secreted from cells. Some researchers use modified cysteines using S-acetomidomethyl (Acm) to block the formation of the disulfide bond but preserve the cysteine and the protein's original primary structure.

Off-resin Cyclization

Off-resin cyclization is a solid-phase synthesis of key intermediates, followed by the key cyclization in solution phase, the final deprotection of any masked side chains is also carried out in solution phase. This has the disadvantages that the efficiencies of solid-phase synthesis are lost in the solution phase steps, that purification from by-products, reagents and unconverted material is required, and that undesired oligomers can be formed if macrocycle formation is involved.

References

- Marquet P, Lachâtre G; Lachâtre (October 1999). "Liquid chromatography-mass spectrometry: potential in forensic and clinical toxicology". Journal of Chromatography B. 733 (1–2): 93–118. PMID 10572976. doi:10.1016/S0378-4347(99)00147-4

- Payne JW (1976). "Peptides and micro-organisms". Advances in Microbial Physiology. Advances in Microbial Physiology. 13: 55–113. ISBN 9780120277131. PMID 775944. doi:10.1016/S0065-2911(08)60038-7

- Ardejani, Maziar S.; Orner, Brendan P. (2013-05-03). "Obey the Peptide Assembly Rules". Science. 340 (6132): 561–562. ISSN 0036-8075. PMID 23641105. doi:10.1126/science.1237708

- Boelsma E, Kloek J; Kloek (March 2009). "Lactotripeptides and antihypertensive effects: a critical review". The British Journal of Nutrition. 101 (6): 776–86. PMID 19061526. doi:10.1017/S0007114508137722

- Webster J, Oxley D; Oxley (2005). "Peptide mass fingerprinting: protein identification using MALDI-TOF mass spectrometry". Methods in Molecular Biology. Methods in Molecular Biology™. 310: 227–40. ISBN 978-1-58829-399-2. PMID 16350956. doi:10.1007/978-1-59259-948-6_16

- Haque, Emily; Chand, Rattan. "Milk protein derived bioactive peptides". Dairy Science. Retrieved 28 July 2014

- Meister A, Anderson ME; Anderson (1983). "Glutathione". Annual Review of Biochemistry. 52 (1): 711–60. PMID 6137189. doi:10.1146/annurev.bi.52.070183.003431

- Bulinski JC (1986). "Peptide antibodies: new tools for cell biology". International Review of Cytology. International Review of Cytology. 103: 281–302. ISBN 9780123645036. PMID 2427468. doi:10.1016/S0074-7696(08)60838-4

- Finking R, Marahiel MA; Marahiel (2004). "Biosynthesis of nonribosomal peptides1". Annual Review of Microbiology. 58 (1): 453–88. PMID 15487945. doi:10.1146/annurev.micro.58.030603.123615

Permissions

Index